16th Russian Youth Conference on Physics of Semiconductors and Nanostructures, Opto- and Nanoelectronics 2014

Journal of Physics: Conference Series Volume 586

St. Petersburg, Russia
24 – 28 November 2014

Editors:

R. A. Suris
L. E. Vorobjev
D. A. Firsov

ISBN: 978-1-63439-916-6
ISSN: 1742-6588

Printed from e-media with permission by:

Curran Associates, Inc.
57 Morehouse Lane
Red Hook, NY 12571

Some format issues inherent in the e-media version may also appear in this print version.

Printed by Curran Associates, Inc. (2015)

For permission requests, please contact the Institute of Physics
at the address below.

Institute of Physics
Dirac House, Temple Back
Bristol BS1 6BE UK

Phone: 44 1 17 929 7481
Fax: 44 1 17 920 0979

techtracking@iop.org

Additional copies of this publication are available from:

Curran Associates, Inc.
57 Morehouse Lane
Red Hook, NY 12571 USA
Phone: 845-758-0400
Fax: 845-758-2634
Email: curran@proceedings.com
Web: www.proceedings.com

TABLE OF CONTENTS

Editorial Preface

The 16th Russian Youth Conference on Physics of Semiconductors and Nanostructures, Opto- and Nanoelectronics was held on November 24 – 28 at St. Petersburg Polytechnic University.

The program of the Conference included semiconductor technology, heterostructures with quantum wells and quantum dots, opto- and nanoelectronic devices, and new materials. A large number of participants with about 200 attendees from many regions of Russia provided a perfect platform for the valuable discussions between students and experienced scientists.

The Conference included two invited talks given by a corresponding member of RAS P.S. Kopyev ("Nitrides: the 4th Nobel Prize on semiconductor heterostructures") and Dr. A.V. Ivanchik ("XXI century is the era of precision cosmology"). Students, graduate and postgraduate students presented their results on plenary and poster sessions. The total number of accepted papers published in Russian (the official conference language) was 92. Here we publish 18 of them in English.

Like previous years, the participants were involved in the competition for the best report. Certificates and cash prizes were awarded to a number of participants for the presentations selected by the Program Committee. Two special E.F. Gross Prizes were given for the best presentations in semiconductor optics. Works with potential applications were recommended for participation in the following competition for support from the Russian Foundation for Assistance to Small Innovative Enterprises in Science and Technology.

The Conference was supported by the Russian Foundation for Basic Research, the "Dynasty" foundation and the innovation company "ATC – Semiconductor Devices", St. Petersburg.

The official Conference website is http://www.semicond.spbstu.ru/conf2014-eng.html

Editors

R.A. Suris
L.E. Vorobjev
D.A. Firsov

Journal of Physics: Conference Series **586** (2015) 011001 doi:10.1088/1742-6596/586/1/011001

RYCPS 2014
XVI RUSSIAN YOUTH CONFERENCE
ON PHYSICS OF SEMICONDUCTORS AND NANOSTRUCTURES, OPTO- AND NANOELECTRONICS
16
November 24-28, 2014
St.Petersburg, Russia

Winners of the E.F. Gross Prize for the best presentation in semiconductor optics

Maxim Eremenko,

a postgraduate student at Ioffe Physical-Technical Institute of the Russian Academy of Sciences, St. Petersburg

Presentation:

"*Exciton states and energy relaxation in low dimensional ZnCdSe with self assembled quantum dots*"

Natalia Kopteva,

a student at St. Petersburg State University

Presentation:

"*Rabi oscillations in a microresonator with a parabolic quantum well*"

Peer review statement

All papers published in this volume of *Journal of Physics: Conference Series* have been peer reviewed through processes administered by the proceedings Editors. Reviews were conducted by expert referees to the professional and scientific standards expected of a proceedings journal published by IOP Publishing.

Mid-infrared light absorption by photo-excited charge carriers in Ge/Si quantum dots

L E Vorobjev[1], D A Firsov[1], V Yu Panevin[1], A N Sofronov[1], R M Balagula[1] and A A Tonkikh[2,3]

[1] St.Petersburg Polytechnic University, Polytechnicheskaya 29, St.Petersburg 195251, Russia

[2] Max Planck Institute of Microstructure Physics, Weinberg 2, Halle (Saale) 06120, Germany

[3] Institute for Physics of Microstructures RAS, GSP-105, Nizhny Novgorod 603950, Russia

e-mail: sofronov@rphf.spbstu.ru

Abstract. Mid-infrared optical absorption spectra of Ge/Si quantum dot structures in polarized light were obtained under conditions of additional interband optical excitation. Photo-excited electron-hole pairs captured to quantum dots cause mid-infrared absorption changes for certain light polarization in spectral range of 0.25 – 0.6 eV. The sign of the effect is found to be different in different spectral ranges. There is an increase of absorption at the long-wavelength edge of the spectrum and a decrease of absorption at the short-wavelength edge. Absorption increase is considered to be related to contribution of optical transitions of non-equilibrium holes from the quantum dot ground states, while absorption decrease is associated with suppression of interband-like processes with generation of hole inside the dot and bound electron outside of the dot under conditions of full occupation of appropriate discrete states.

1. Introduction

Ge/Si quantum dot structures are attractive objects for a number of scientific groups around the world. Intensive studies of optical and photoelectric properties of Ge/Si quantum dots are explained by a possibility to develop the mid- and near-infrared optoelectronic devices based on these structures. One of the proposed advantages of this material system is the possibility of integration into the existing silicon technology.

In the recent several years a noticeable progress has been made in technology aspects of Ge/Si quantum dots growth (see, for example, [1] and reference therein), in theoretical studies of specific charge carriers quantization features [2] and experimental studies of near-infrared [3, 4] and mid-infrared [5, 6] optical properties, and in device development [7].

The present work is focused on the optical properties of Ge/Si quantum dots in the mid-infrared spectral range under conditions of non-equilibrium population of the dot states with photo-excited charge carriers.

2. Experimental setup and samples

Structure with self-assembled MBE-grown Ge/Si quantum dots on Si (001) substrate was studied. Structure contains 10 periods of quantum dot layers formed by deposition of 7 Ge monolayers with 0.14 ML/s rate at 600^{0}C. Quantum dot layers were separated from each other with 15 nm Si interlayer. The structure was separated from substrate with an epitaxial 100 nm silicon buffer layer and was covered with a 100 nm cover layer (see figure 1(a)). A surfactant (Sb) was used during the growth process in order to increase the density of the QD array as described in [8].

The form and dimensions of QDs were estimated using transmission electron microscopy. Quantum dots have a pyramidal shape with the average height of 2.7 nm, and the average base size of 14 nm. The results of atomic force microscopy of specially grown samples with one QD layer and no surface silicon layer showed that the QD density was $2 \cdot 10^{11}$ cm^{-2}. The approximate content of germanium in a resulting quantum dot material is 60–65%. No special doping was performed during structure growth.

The samples in common multi-pass geometry were prepared from the structure for optical measurements in the mid-infrared spectral range (see figure 1(b)). Two opposite sample sides were mechanically processed so the angle between the incident beam (normal to the side) and the top surface of the sample was 45°. The light beam passed the structure multiple times due to internal reflection in the sample. This geometry also allows us to use different light polarization with respect to growth direction.

Since the structure under investigation was not specially doped, the charge carriers were created only using additional illumination of the top surface of the sample with a solid-state 532 nm laser.

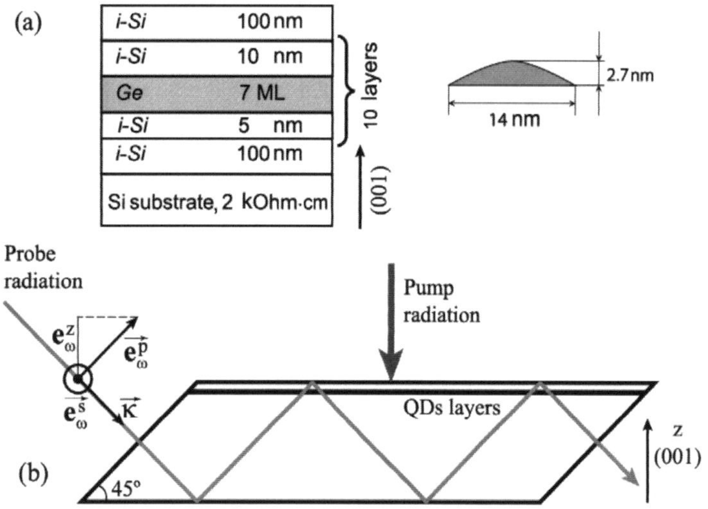

Figure 1. Schematics of the structure, dot shape (a), and experimental sample geometry (b). Vector diagram shows the orientation of probe radiation wave vector $\vec{K}$ and two possible polarization vectors $\vec{e}_\omega$ for *s*- and *p*-polarization. The last one has the non-zero component in the direction *z* of growth axis.

Experiment scheme in general is similar to the variation of steady state pump-probe technique. The absorption spectra in the mid-infrared range were measured with an automated experimental setup based on a prismatic monochromator with a globar as a source of broadband infrared radiation. A spherical mirror based optical scheme formed a thin strip of monochromatic infrared light (probe radiation) at the canted side surface of the sample and focused light from the opposite side of the sample to the photosensitive area of liquid nitrogen cooled InSb photodetector. Probe radiation polarization was selected with a metal grid ZnSe polarizer. The sample was mounted in a liquid nitrogen cryostat with a pair of ZnSe optical windows in the direction of probe light propagation and a quartz optical window in the orthogonal direction to allow the excitation of sample with interband laser radiation (pump radiation). Probe radiation was mechanically chopped at 900 Hz, simultaneously exciting laser radiation was chopped at 90 Hz. The photodetector signal was measured simultaneously with two SR830 Lock-In amplifiers. The first one was locked to the frequency of infrared radiation modulation (900 Hz) and measured a signal proportional to the equilibrium optical transmittance T of the sample. The second one was locked to the excitation modulation frequency (90 Hz) and measured a signal proportional to the variation of the optical transmittance ΔT, related only to non-equilibrium photo-excited charge carriers.

3. Experimental results

It is well known that Ge/Si material system forms heterointerfaces of second type with a quantum well for holes in the valence band and a barrier for electrons in the conduction band. A pump radiation generates non-equilibrium electron-hole pairs with a relatively large energy. After fast relaxation processes the holes are captured inside the quantum dots, while the electrons tend to be localized around the dot as discussed later in this section. This non-equilibrium population of quantum dot energy spectrum states can effect the optical absorption of the probe radiation. A complicated shape of the quantum dot and a noticeable size difference of the dot in the plane of structure and in the growth direction allow one to expect a certain selection rules of the intraband absorption for in-plane *xy*-polarization and *z*-polarization along the growth direction. In experiment, we measured $\Delta T/T$ spectra for *s*- and *p*-polarization of probing radiation. The typical measured spectra of photo-induced absorption for these two polarizations are shown in figure 2.

Since the polarization vector for *p*-polarization in experiment contains equal *xy*- and *z*-components (while *s*-polarization is in fact *xy*- one, see figure 1(b)), one can obtain the value of $\Delta T/T$ for *z*-polarized light with simple conversion $(\Delta T/T)_z = 2((\Delta T/T)_p - (\Delta T/T)_s /2)$. The resulted spectra are shown in figure 3.

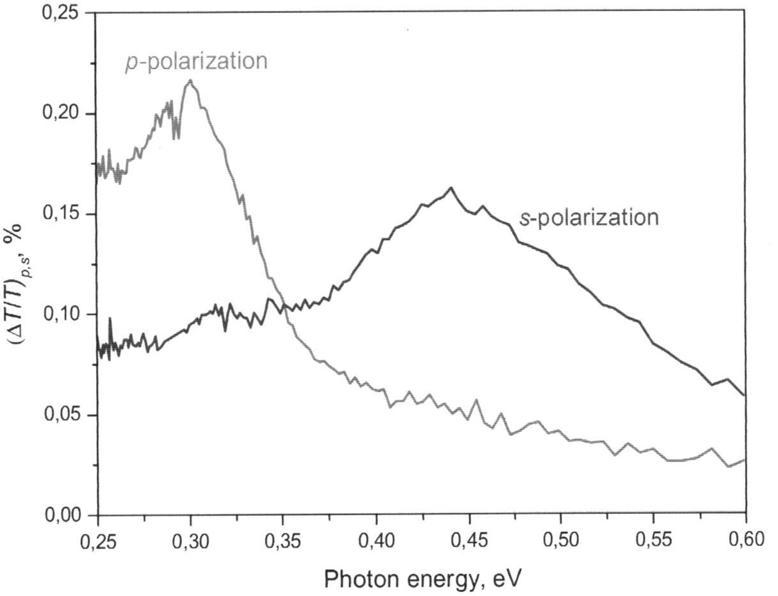

Figure 2. Measured spectra of photo-induced absorption for *p*- and *s*- infrared light polarization at $T = 79$ K and pump intensity $I = 20$ mW.

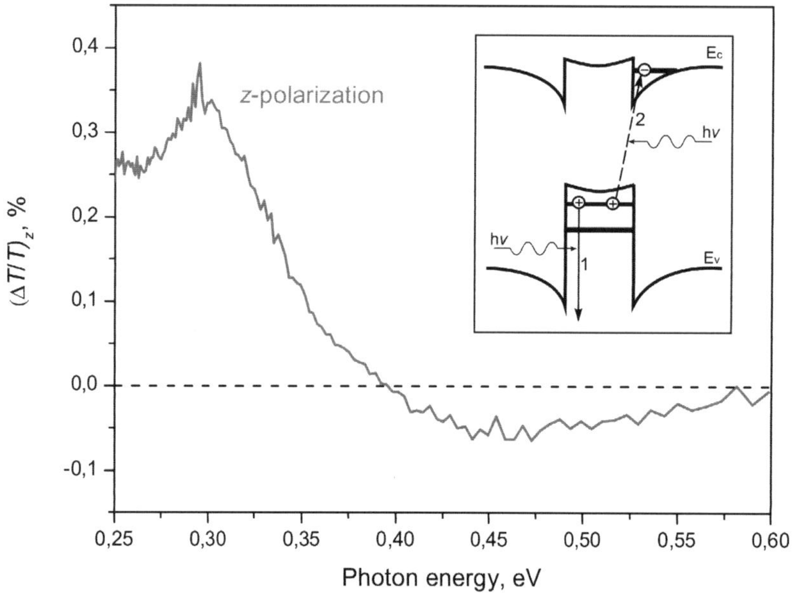

Figure 3. Spectrum of photo-induced absorption for infrared light polarized along the growth direction calculated from measured data at $T = 79$ K and pump intensity $I = 20$ mW. Inset shows the optical transitions scheme. Transitions marked with solid arrow 1 are responsible for absorption increase at long-wavelength part of spectrum. Suppression of transitions marked with dashed arrow 2 is responsible for absorption change drop below zero.

One can see that presence of nonequilibrium charge carriers leads to an increase in the optical absorption for z-polarized light in a wide spectral range with the centre at 0.3 eV and to a decrease in the absorption at photon energies about 0.45 eV. The absorption at 0.3 eV is related to hole optical transitions from the ground state of quantum dot to continuum of states above the barrier (see inset on figure 3, where this transition is depicted with solid arrow 1). The spectral position of this absorption feature correlates well with the previously measured mid-infrared photoconductivity spectra in similar doped structure [9]. Inhomogeneous broadening of observed absorption peaks is related to the variation of quantum dots geometry.

The decrease of absorption at 0.45 eV can be related to the dynamic analogue of bulk Burstein–Moss effect. Processes of photogeneration of electrons around the dot and holes inside the dot by probe radiation become impossible due to the full population of the final states for these transitions by charge carriers generated by pump radiation. This type of transition is shown at inset in figure 3 with dashed arrow 2. Again, the spectral position of this effect is in good agreement with the results of interband photoconductivity measurements in Ge/Si quantum dot structures [4].

The observed polarization of the interband absorption can be related to the complicated potential profile of the silicon conduction band at the quantum dot heterointerface. In particular, non-uniform distribution of mechanical strain in a multilayer structure leads to the energy shift of only two X-valleys of conduction band aligned along the structure growth direction, while the other four valleys remain unshifted [2]. This leads to the angle-dependent potential profile, and electron localization arises only at certain side of the dot, but not uniformly around the dot.

The temperature dependence of the $\Delta T/T$ value at the spectral maximum was also measured. The experimental data demonstrate a significant suppression of the observed effect with the increase of temperature. At the temperatures above 250 K the variation of the optical transmittance related to non-equilibrium charge carriers becomes negligible.

4. Conclusion

In conclusion, in this work the photo-induced mid-infrared absorption spectra were measured in Ge/Si quantum dot structures. An increase of absorption under interband optical excitation related to the contribution of optical transitions of non-equilibrium holes from the quantum dot ground state is observed for light polarized along the growth direction. At the same time, at shorter wavelengths the decrease of absorption is observed for the same light polarization. This effect is considered to be caused by the suppression of interband-like optical transitions between bound states in the valence and conduction bands and can be described as a dynamic analogue of the Burstein–Moss effect.

Acknowledgments

This work has been supported by the Russian Foundation for Basic Research (Grant № 13-02-12203) and the Ministry of Education and Science of the Russian Federation.

References

[1] Tonkikh A A, Zakharov N D, Novikov AV, Kudryavtsev KE, Talalaev V G, Fuhrmann B, Leipner H S and Werner P 2012 *Thin Solid Films* **520** 3322
[2] Yakimov A I, Dvurechenskii A V, Nikiforov A I, Bloshkin A A, Nenashev A V and Volodin V A 2006 *Phys. Rev.* B **73** 115333
[3] Schilling J, Talalaev V, Tonkikh A, Fuhrmann B, Heyroth F and Otto M 2013 *Appl. Phys. Lett.* **103** 161106
[4] Talochkin A B and Chistokhin I B 2011 *Semiconductors* **45** 907
[5] Yakimov A I, Kirienko V V, Armbrister V A, Bloshkin A A and Dvurechenskii A V 2014 *Phys. Rev.* B **90** 035430
[6] Vorobjev L E, Firsov D A, Shalygin V A, Panevin V Yu, Sofronov A N, Yakimov A I, Dvurechenskii A V, Tonkikh A A and Werner P 2012 *Semiconductors* **46** 1529
[7] Yakimov A I, Dvurechenskii A V, Nikiforov A I and Proskuryakov Yu Yu 2001 *J. Appl. Phys.* **89** 5676
[8] Tonkikh A, Zakharov N, Talalaev V and Werner P 2010 *Phys. Status Solidi RRL* **4** 224
[9] Panevin V Yu, *et al* 2013 *Semiconductors* **47** 1574

X-ray radiation influence on photoluminescence spectra of composite thin films based on C_{60}<CdTe>

M A Elistratova[1,2], I B Zakharova[1] and N M Romanov[1,2]

[1] St. Petersburg Polytechnic University, 29 Polytechnicheskaya str., St. Petersburg 195251, Russia
[2] Lappeenranta University of Technology, 34 Skinnarilankatu, Lappeenranta 53850, Finland

E-mail: elistratovamari@yandex.ru

Abstract. Photoluminescence spectra of composite thin films based on C_{60}<CdTe> before and after X-ray irradiation, as well as the results of quantum-chemical calculations of corresponding molecular complexes are presented. Fullerene films doped by CdTe with various concentrations were obtained by means of vacuum co-evaporation in a Knudsen cell. Composition and surface morphology were measured by secondary electron microscopy and energy-dispersive X-ray spectroscopy. X-ray irradiated films were considered, and additional peaks in photoluminescence spectra were detected. These peaks appear as a result of molecular complexes formation from C_{60}CdTe mixture and dimerization of the films. Density functional B3LYP quantum-chemical calculations for C_{60}CdTe, molecular complexes, $(C_{60})_2$ and $C_{120}O$ dimers were performed to elucidate some experimental results.

Introduction

Application of nanostructured materials in light energy conversion has emerged as an alternative to single-crystalline devices. Low material cost, easy methods of fabrication, high light absorption cross sections and ability to tune the optical response make organic-inorganic nanocomposites attractive candidates for optoelectronic devices. To achieve effective photoinduced charge separation, the bulk heterojunction can be used. To produce the bulk heterojunction the continuous interpenetrating network of donor and acceptor semiconductor is required. Nanoscale carbon, such as nanotubes and fullerene, represents a class of materials that are actively explored for solar energy applications [1]. For example, hybrid organic/inorganic composites combining C_{60} with semiconductor CdSe quantum dots (QDs) were synthesized and tested for light-energy harvesting [2]. Due to the strong acceptor properties fullerene and its modifications are widely used to produce the molecular complexes, and various organic and inorganic materials are used as a donor. In the same way, A^2B^6 group is of particular interest for their application in solar cell creation. Valuable information about the possibility of creation of new fullerene based composite materials can be obtained with use of ab initio quantum-chemical calculation [3]. It should be mentioned that numerous studies [4, 5] are dedicated to the structure, optical properties and the nature of various fullerene-inorganic clusters at the molecular and supramolecular level. However, irradiation effect on the material property is still out of sight of researchers. Therefore, the choice of the subject matter of this study is relevant from both the scientific and applied perspectives. We report the results of a study of photoluminescence spectra of the thin composite C_{60}<CdTe> films before and after X-ray irradiation, as well as the results of quantum-chemical calculations of corresponding molecular complexes.

Sample preparation and experimental methods

Thin films of $(C_{60})_{1-x}(CdTe)_x$ were prepared by co-deposition technique from a Knudsen cell or quasi-closed volume. A possibility of large area film fabrication is an advantage of the Knudsen cell. The mixture of C_{60} with 10-50 wt.% of CdTe was evaporated at 520–650 °C on Si, a glass with ITO, KBr or mica substrate. This process may give rise to a significant changing of a film composition in comparison with the initial charge. Changes in the composition are much less in the case of quasi-closed volume evaporator than in the case of the Knudsen cell. The substrates were kept at room temperature during the process. Sputtering on a cold substrate leads to amorphous and polycrystalline structure formation and possibly to alignment of randomly distributed semiconductor clusters or molecular complexes. A scanning electron microscope Jeol JSM-6390 with resolution of 3 nm was used to study the surface morphology. The films composition in the selected area was measured by an energy dispersive micro-analysis console "Oxford INCA Energy" with the utmost sensitivity of 0.1 wt%. These studies confirmed the high homogeneity of the impurity distribution and the absence of the mosaic structure of the composite films.

The samples were subjected to X-ray irradiation in REIS-D equipment with a rhenium anode. X-ray energy flux was $5 \cdot 10^{-3}$ J/s during 6 hour. Photoluminescence (PL) spectra were obtained by automated equipment based on a FHR 640 monochromator with a grating of 1200 mm^{-1} and the Symphony II (1024×256) Cryogenic Open - Electrode CCD detector. Samples were fixed at the angle of 45° to the laser beam. A quarts cryostat was used to make measurements at 77 K in liquid nitrogen.

Results and discussion

PL spectra obtained at 300 K are shown in figure 1. The wide PL peak with 730 nm maximum is detected in pristine fullerene films. Since singlet transitions for the C_{60} molecule are prohibited for the

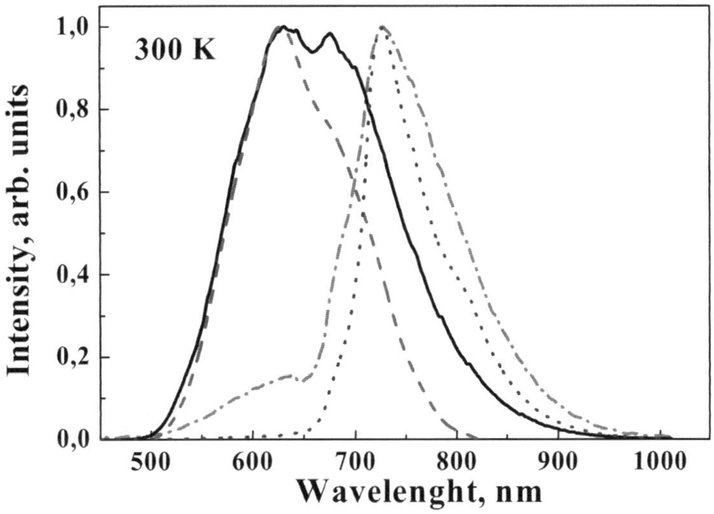

Figure 1. Photoluminescence spectra obtained at 300 K. X-ray irradiated pure C_{60} spectrum is shown by dash line. X-ray irradiated C_{60} with 50% CdTe is shown by solid line. Unexposed C_{60} with 50% CdTe and pure C_{60} PL spectra is dash-dot and dot line accordingly.

reason of symmetry, the wide peak can be interpreted as the superposition of two fullerene peaks: T_1-T_0 at 730 nm and its vibration recurrence at 820 nm.

On addition of CdTe to fullerene the extra peak appears at the wavelength of 630 nm. The extra peak intensity increases with impurity content. This peak may be attributed to the emergence of singlet excitation in C_{60} electron spectrum. The corresponding optical transition is forbidden for a fullerene molecule in fullerite at room temperature for the reason of symmetry (T_h point group). However, CdTe doping leads to molecular complex formation and reduces the symmetry of C_{60}, so that the transition becomes allowed. The changes in PL spectra for irradiated CdTe-doped and pristine C_{60} films were investigated. The results have shown significant changes in the photoluminescence spectra after X-ray irradiation dose. A wide peak is observed only at 600−650 nm in the case of exposed samples. Therefore, after X-ray exposition the wide peak (T_1−T_0) at 730 nm disappears and another wide peak (S_1−S_0) emerges at 630 nm. T_1−T_0 is a transition between triplet fullerene levels and S_1−S_0 is a transition between singlet levels. More detailed information on the electronic structure comes from PL measurements at 77 K (see figure 2). One can see the weak peak at 630 nm in pristine fullerene films. The corresponding optical transition appears because the symmetry of fullerite is reduced at temperatures below 260 K and 90 K and a simple cubic lattice and glass phase are respectively formed.

All additional peaks mentioned above are detected in X-ray exposed pristine fullerene. The energy position of the peaks does not change significantly but for 630 and 680 nm peaks the relative intensity increases after irradiation. In the case of composite films the changes are significantly stronger and initial emission peaks (730 nm and 820 nm) almost disappear. To explain the results, one can suggest that X-ray irradiation promotes dimerization in the fullerene matrix and reduces the symmetry, so that some optical transitions become possible.

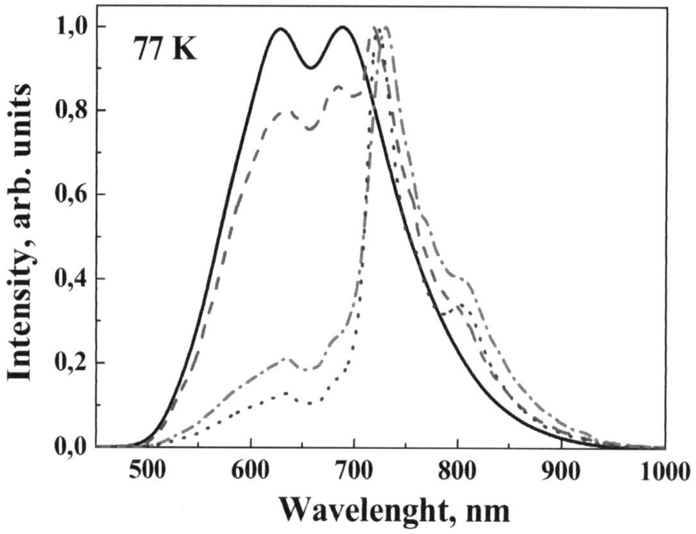

Figure 2. Photoluminescence spectra obtained at 77 K. X-ray irradiated pure C_{60} spectrum is shown by dash line. X-ray irradiated C_{60} with 50% CdTe is shown by solid line. Unexposed C_{60} with 50% CdTe and pure C_{60} PL spectra is dash-dot and dot line accordingly.

Quantum chemical calculations of the optimum geometry, total energy and electronic structure of $C_{60}CdTe$ molecular complex, CA (cyclo-adduct) and SB (single-bonded) dimers $(C_{60})_2$ and $(C_{60})_2O$

were performed by the DFT-B3LYP method. It is shown that in electronic spectra of dimers and complexes allowed singlet-singlet transitions of about 1.91 eV appear in accordance with the experimental data.

Conclusions

The novel method for the production of thin composite fullerene films was developed. The samples were investigated using the optical spectroscopy technique. In this work it was revealed that X-ray irradiation affects significantly the electronic structure of both pristine fullerene films and the films doped by CdTe. The photoluminescence spectra measured have shown the appearance of some extra peaks related to the supramolecular complex formation. It was revealed that the 630 nm (1.97 eV) line appears in the doped films, which is absent in pure fullerene. Quantum-chemical calculations have shown that this line may be attributed to S_1–S_0 transition in electron excitation spectrum which is forbidden in pristine fullerene. It is found that intensive X-ray irradiation of the films gives rise to considerable enhancement of intensity of the additional peak which becomes prevalent in spectra of composite films. The possible reason is that X-ray irradiation brings to dimerization of the C_{60} matrix.

References

[1] Shibu E S *et al* 2012 *ACS Nano* **6** 1601
[2] Bang J H and Kamat P V 2011 *ACS Nano* **5** 9421
[3] Chaban V V, Prezhdo V V and Prezhdo O V 2013 *J. Phys. Chem. Lett.* **4** 1
[4] Guldi D M, Zilbermann I, Anderson G, Kotov N A, Tagmatarchis N and Prato M 2004 *J. Am. Chem. Soc.* **126** 14340
[5] Brown P and Kamat P V 2008 *J. Am. Chem. Soc.* **130** 8890

The use of asymptotic methods for modelling of the carriers wave functions in the Si/SiGe heterostructures with quantum-confined layers

A Orlov[1], N Levashova[1] and T Burbaev[2]

[1]Moscow State University, Faculty of Physics, Leninskie Gory 1-2, Moscow 119991, Russia
[2]Lebedev Physical Institute of the Russian Academy of Sciences, 53 Leninskiy Prospect, Moscow 119991, Russia

E-mail: orlov.andrey@physics.msu.ru

Abstract. The mathematical model that allows obtaining wave functions of electrons and holes and Coulomb potential in dependence on the parameters of the layer is proposed. In contrast to the common approach widely used in considerable literature, the proposed method allows to find the wave functions for all layers of the structure, both inside and outside the barrier layer. It was shown that increase of tunnel barrier transparency entails a transition from the dipolar electron-hole system (EHS) with a double-peak wave function of electrons to the spatially direct EHS. The proposed model takes into account the effect of the correlation interaction on the wave functions and the density of distribution of the carriers crosswise the layers. These calculations are useful for the analysis of many-particle nonequilibrium states occurring in low-dimensional structures.

1. Introduction

The pioneer works [1, 2] have laid the foundation of the wide investigations in physics of many-particle interaction concerning low-dimensional exciton and electron-hole systems (EHS) of high density. The research was mainly held in the type-I heterostructures GaAs/AlGaAs with quantum wells (QW) both for electrons and holes. In case of high density of nonequilibrium charge carriers there were detected different phases of the condensed matter – the dipolar electron-hole liquid, the Bose-Einstein matter with challenging coherent, superconductive, and superfluid properties. In the present work we investigated the type-II $Si/Si_{1-x}Ge_x/Si$ heterostructures with a strained layer $Si_{1-x}Ge_x$ forming a rather deep potential well for holes and not a high barrier for electrons in the zone diagram. The height of the barrier grows with increasing content of germanium x [3, 4]. In such structures the EHS must have not less interesting properties [5]. In the work the behavior of the wave functions of charge carriers when changing the barrier parameters is modelled. Consider the Bohr radius of the exciton in bulk Si to be about 5 nm; the calculations were held for heterostructures with the thickness of solid solution L varying from 1 to 70 nm. The distribution of the carrier density crosswise the layer is obtained by solving the system of three equations: the two one-particle Schrödinger equations both for holes and electrons and the Poisson equation for the Coulomb energy of their interaction. Commonly in considerable literature the solution of such a system is obtained separately outside and inside the layer and then is joined smoothly at the quantum barrier borders. The correlative interaction of the carriers influential in the case of high density is then neglected. In the present work the correlative interaction was counted according to the method offered in [6], then the asymptotic analysis was held and the numerical solution was obtained using the "hard" one-staged complex-valued Rosenbrock scheme. We didn't employ any additional technique of smooth joining. The numerical calculations showed that with decreasing height and/or width of the barrier the transition from the spatially indirect (dipolar) EHS to the spatially direct one occurs. Also it was revealed that there are some characteristics of the barrier in the case of high carrier density when the combined formation of both spatially direct and dipolar states is possible.

2. The model

To describe the form of the charge carriers wave functions in quantum-confined layers in the type-II heterostructure ($Si/Si_{1-x}Ge_x/Si$) a one-dimensional boundary value problem for a system of three equations is set: two one-particle Schrödinger equations and the Poisson equation for the Coulomb energy:

$$\begin{cases} \frac{\hbar^2}{2m^*}\frac{d^2\Psi e}{dz^2} = \left(V_e(z) - \varphi(z) - E_e\right)\Psi e + E_{Ce}\Psi e \\ \frac{\hbar^2}{2m^*}\frac{d^2\Psi h}{dz^2} = \left(V_h(z) + \varphi(z) - E_h\right)\Psi h + E_{Ch}\Psi h \\ \frac{d^2\varphi}{dz^2} = \frac{e^2 n_0}{\varepsilon\varepsilon_0}\left(|\Psi e|^2 - |\Psi h|^2\right) \end{cases} \tag{1}$$

Here $\hbar$ is the Planck constant, ε – the dielectric constant, ε_0 – the electric constant, m^* is the effective mass of the carrier; it is intended that the effective mass of the electron, m_e^* and the effective mass of the hole, m_h^* are equal: $m^* = m_e^* = m_h^*$; e is the electron charge; n_0 is the sheet carrier density appearing in the system due to external radiation; Ψe, Ψh are normalized to unity wave functions of electrons and holes respectively: $\int_{-\infty}^{+\infty} (\Psi e)^2\, dz = 1$; . $\int_{-\infty}^{+\infty} (\Psi h)^2\, dz = 1$; φ is the Coulomb potential energy. The potential barrier for electrons $V_e(z)$ and the potential well for holes $V_h(z)$ are defined as following:

$$V_e(z) = \begin{cases} 0, & |z| > \frac{L}{2}, \\ V_e, & |z| < \frac{L}{2}; \end{cases} \qquad V_h(z) = \begin{cases} 0, & |z| > \frac{L}{2}, \\ -V_h, & |z| < \frac{L}{2}. \end{cases} \tag{2}$$

Here L is the width of the layer in nm.

In the model the correlative interaction was counted according to the method of density functional theory [6]. For this purpose, the components $E_{Ce}\Psi e$ and $E_{Ch}\Psi h$ were introduced in the Schrödinger equations in the system (1). As it was shown in [8] the correlation energy of the carriers is related to their wave functions $\Psi_{carrier}$ in the following way: $E_C \sim (n_0)^{1/4} \sim (\Psi_{carrier})^{1/2}$. The system (1) is considered on the segment $-A \le z \le A$, where A is a value greater than the strained layer width. On the borders the homogeneous Neumann conditions are adjusted. We rewrite the system in dimensionless form. We choose the exciton energy $E_x = \frac{me^4}{(4\pi\varepsilon_0)^2\,\varepsilon^2\hbar^2}$ as a scale energy and the value $\frac{a_x}{\mu}$ as a scale length. Here $a_x = \frac{4\pi\varepsilon_0\varepsilon\hbar^2}{me^2}$ is the Bohr radius of the exciton, m is defined from a ratio $\frac{1}{m} = \frac{1}{m_e^*} + \frac{1}{m_h^*}$; $\mu \sim 0.01$ is a small parameter introduced in the system artificially to obtain the solutions including the interior transition layers.

In the dimensionless form the boundary value problem becomes the following:

$$\begin{cases} \mu^2 \dfrac{d^2\Psi e}{dz^2} = \left(V_e(z) - \varphi(z) - E_e\right)\Psi e + N|\Psi e|^{1/2}\,\Psi e \\ \mu^2 \dfrac{d^2\Psi h}{dz^2} = \left(V_h(z) + \varphi(z) - E_h\right)\Psi h + N|\Psi h|^{1/2}\,\Psi h \\ \dfrac{d^2\varphi}{dz^2} = K\left(|\Psi e|^2 - |\Psi h|^2\right). \end{cases} \tag{3}$$

Boundary conditions are the following:

$$\left.\frac{d\Psi e}{dz}\right|_{z=-\frac{\mu A}{a_x}} = \left.\frac{d\Psi e}{dz}\right|_{z=\frac{\mu A}{a_x}} = \left.\frac{d\Psi h}{dz}\right|_{z=-\frac{\mu A}{a_x}} = \left.\frac{d\Psi h}{dz}\right|_{z=\frac{\mu A}{a_x}} = 0; \quad \left.\frac{d\varphi}{dz}\right|_{z=-\frac{\mu A}{a_x}} = \left.\frac{d\varphi}{dz}\right|_{z=\frac{\mu A}{a_x}} = 0. \tag{4}$$

Here N – is a constant chosen in the course of numerical experiments that depends on the strained layer characteristics, $K = 4\pi \frac{n_0 \cdot a_x^2}{\mu} \sim 1$.

3. Asymptotic analysis

The research procedure of the systems of equations with a small parameter besides the derivatives is proposed in [7]. If we put $\mu = 0$ in the first two equations of the system (3), we obtain the so-called degenerate system:

$$\begin{cases} \left(V_e(z) - \bar{\varphi}(z) - E_e\right)\bar{\Psi}e + N|\bar{\Psi}e|^{1/2}\,\bar{\Psi}e = 0 \\ \left(V_h(z) + \bar{\varphi}(z) - E_h\right)\bar{\Psi}h + N|\bar{\Psi}h|^{1/2}\,\bar{\Psi}h = 0 \end{cases} \tag{5}$$

The numerical solution of the system (3) is attracted to a stable solution of the degenerate system at a sufficient distance from the borders of the layer. The second equation (5) has two roots: $\bar{\Psi}h = (V_h(z) - \bar{\varphi}(z) + E_h)^2 \cdot N^{-2}$ and $\bar{\Psi}h = 0$. Outside the strained layer where $V_h(z) = 0$ the stable root is $\bar{\Psi}h^{(ex)} = 0$ and inside the layer the stable root is $\bar{\Psi}h^{(in)} = (V_h - \bar{\varphi}(z) + E_h)^2 \cdot N^{-2}$, i.e. the holes are concentrated within the layer. Outside the layer the stable root of the first equation (5) is $\bar{\Psi}e^{(ex)} = \left(\bar{\varphi}(z) + E_e\right)^2 \cdot N^{-2}$, that corresponds to the concentration of electrons outside the strained layer. Inside the layer where the inequality $V_e \geq \bar{\varphi}(z) + E_e$ is held the first equation (5) has a unique solution $\bar{\Psi}e^{(in)} = 0$ (the electrons are ejected from the layer). If $V_e < \bar{\varphi}(z) + E_e$ the zero root becomes unstable and there is the second, non-zero, root of the equation $\bar{\Psi}e^{(in)} = \left(-V_e + \bar{\varphi}(z) + E_e\right)^2 \cdot N^{-2}$, which is stable in this case. This means that at sufficiently low height of the barrier the electrons can concentrate within the layer.

The potential energy in the zeroth approximation (in powers of the small parameter μ) can be determined from the third equation of system (3), with substituted squares of the wave functions $\bar{\Psi}e$ and $\bar{\Psi}h$ derived from a degenerate system. Using the zeroth approximation for the potential, we can determine the energy level for electrons E_e. In the present work we used the method of inverse iterations for this purpose. The holes energy level, E_h is obtained using the common technique for calculation of energy levels for a particle in a potential well.

4. The results of numerical experiments

The results of numerical experiments are shown in figure 1. In the figure the normalized density distribution of electrons (solid line) and holes (dashed line) crosswise the layer is represented. Vertical dashed lines indicate the boundaries of the SiGe layer. The main conclusions are the following.

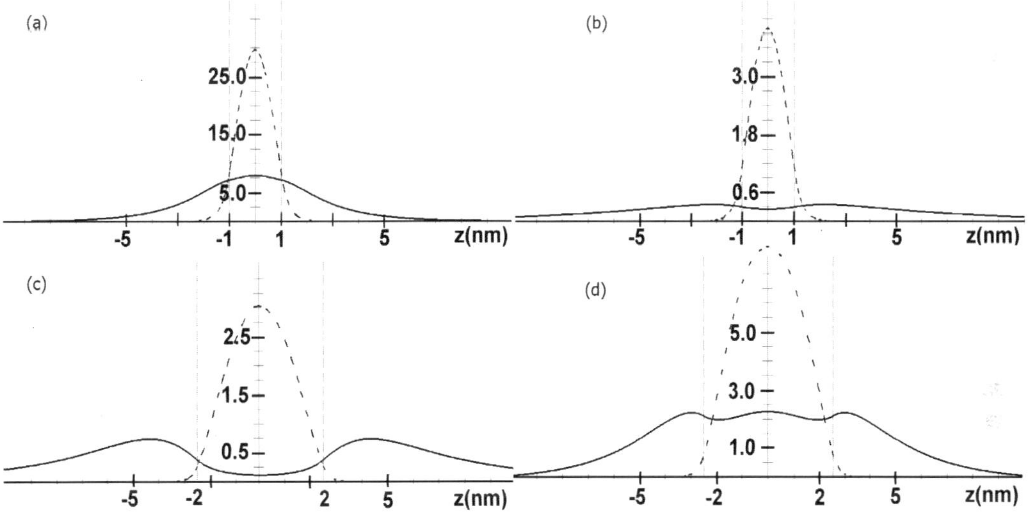

Figure1. Normalized density distribution of electrons (solid line) and holes (dashed line) crosswise the layer SiGe at different values of the layer width L, the barrier height V_e. and the parameter N. Vertical dashed lines indicate the boundaries of the SiGe layer.
(a) $L = 2$ nm, $V_e = 4$ meV, $N = 0.09$; (b) $L = 2$ nm, $V_e = 5$ meV, $N = 0.18$;
(c) $L = 5$ nm, $V_e = 10$ meV, $N = 0.081$; (d) $L = 5$ nm, $V_e = 10$ meV, $N = 0.055$.

At low barrier height ($V_e \leq 5$ meV, figure 1 (a)) and the arbitrary layer width the spatially direct EHS is implemented. Despite the barrier in the conduction band the maximum of the electron density is located in the center of the SiGe layer. (The wave function for holes in all cases is well localized in QW formed by SiGe layer in the valence band, however, the distribution of the electron and hole densities crosswise the layer does not coincide). This behavior of the electron wave function is qualitatively explained in [9] by the fact that for the concentrations of nonequilibrium carriers commonly used in the experiments the band bending caused by the Coulomb interaction of spatially separated electrons and holes creates a potential well for electrons in the SiGe layer (Hartree potential) comparable with the barrier height, in fact transforming these structures into a type-I heterostructure.

Note, that in such structures (at $x \approx 0.05$) for the layer width of 70 and 5 nm, respectively, the electron-hole liquid (EHL) was observed, which was spatially direct [10, 11].

With the increase of the barrier height (V_e = 5 meV, figure 1 (b)) the distribution of electrons crosswise the layer becomes double-peak. The transition to the dipolar EHS occurs. With further increase of the width and/or height of the barrier (V_e = 10 meV, L = 5 nm, figure 1 (c)) and not very high carrier density (low correlation interaction) the EHS becomes dipolar. Exactly in such structures upon the excitation the dipolar excitons and biexcitons must occur, and at high concentrations of nonequilibrium carriers the dipolar EHL must appear.

In the structure with the same parameters (figure 1 (d)) but higher correlative interaction (i.e. with higher density of nonequilibrium carriers) the two-component EHS is implemented. The holes are still well localized in the SiGe layer. The electron wave function has three maxima: two symmetric peaks in the Si layer and a less intense maximum at the center of the layer. In such a system the observation of the two-component excited state is possible.

5. Conclusion

The mathematical model that allows obtaining wave functions of electrons and holes and Coulomb potential in dependence on the parameters of the layer is proposed. In contrast to the common approach widely used in considerable literature, the proposed method allows to find the wave functions (and therefore the distribution of the carriers crosswise the layer) for all layers of the structure, both inside and outside the barrier layer, and does not require any additional artificial technique of smooth joining of the wave functions at the boundaries of the layers. The calculations showed that with decreasing barrier height and/or width the transition from the spatially indirect (dipolar) EHS to the spatially direct one occurs. The proposed model takes into account the effect of the correlation interaction on the wave functions and the density of distribution of the carriers crosswise the layers. This consideration is significant at high density of nonequilibrium carriers (i.e. at high level of excitation). It is shown that under certain parameters of the barrier at high carrier density the formation of both spatially direct and dipolar states is possible.

Acknowledgments

This study was supported by grants of the Russian Foundation for Basic Research (RFBR 13-02-00853 and 13–01–00200).

References

[1] Larionov A V and Timofeev V B 2001 *JETP Lett.* **73** 301
[2] Butov L V, Gossard A C and Chemla D S 2002 *Nature* **418** 751
[3] Van de Walle C G and Martin R M 1996 *Phys. Rev.* B **34** 5621
[4] Rieger M M and Vogl P 1993 *Phys. Rev.* B **48** 14276
[5] Burbaev T M, Kozyrev D S, Sibeldin N N and Skorikov M L 2013 *JETP Lett.* **98** 823
[6] Kohn W and Sham L J 1965 *Phys. Rev.* B **140** A 1133
[7] Butuzov V F, Levashova N T and Mel'nikova A A 2013 *CMMP* **53** N9
[8] Silin A P 1983 *Kratkie Soobshcheniya po Fizike, in Russian* **5** 30
[9] Thewalt M L W, Harrison D A, Reinhart C F and Wolk J A 1997 *Phys. Rev. Lett.* **79** 269
[10] Burbaev T M, Bobrik E A, Kurbatov V A, Rzaev M M Sibel'din N N, Tsverkov V A and Schäffler F 2007 *JETP Lett.* **85** 331
[11] Bagaev V S, Krivobok V S, Nikolaev S N, Novikov A V, Onishchenko E E and Skorikov M L 2010 *Phys. Rev.* B **82** 1153131

Features of microstructure of InSb quantum dots on InAs substrate

N A Bert, V N Nevedomskiy and L A Sokura[1]

Ioffe Institute, 26 Polytekhnicheskaya, 194021 St. Petersburg, Russian Federation

[1]e-mail: sokura@mail.ioffe.ru

Abstract. The properties of InSb/InAs quantum dots (QDs) were investigated by transmission electron microscopy (TEM). Specific features of diffraction contrast were discovered in plan-view TEM images of big (9–10 nm in height and 38–50 nm in diameter) InSb QDs. The simulation of image contrast from the InSb/InAs QD containing a partial Frank dislocation (FD) was carried out for the first time. The comparison of experimental and simulated data allowed us to explain the observed features of the moiré pattern in the image of big InSb QD by the presence of a misfit defect (FD) at the QD-substrate interface.

1. Introduction

Heterostructures with InSb QDs on GaSb or InAs substrates are potential for producing optoelectronic devices operating in the mid-infrared range (3–5 μm) [1]. For instance, mid-IR lasers are used in various fields – chemical process control, environmental monitoring, non-invasive medical diagnostics, leak detection and laser surgery.

The structural properties of InSb/InAs system are similar to those of the well studied InAs/GaAs system. Its crystal structure also forms the face-centered cubic (fcc) lattice. The InSb/InAs system has the lattice mismatch of 6.9 % which is quite close to that in InAs/GaAs (7.2 %). However, because the conditions of InSb/InAs QD formation are not optimized yet the big islands appear among small ones during the epitaxial growth of InSb on InAs substrate [2–3]. Such big dots are incoherent and contain misfit defects.

The subject under study was determination of misfit defect nature in big incoherent InSb/InAs QDs. For this purpose the microstructure of InSb/InAs QDs was studied by TEM. Based on TEM results the model of the structure of such QDs was developed, and calculations of displacement fields and modeling of diffraction contrast TEM images were done.

2. Experimental

The samples were grown using the LPE or MOCVD technique. The LPE growth was carried out in a horizontal set-up equipped with a conventional graphite slider boat under purified H_2 flow on InAs(001) substrate in the temperature range of 420–445°C during 2 seconds, the temperature was lowered from the saturation point down to the growth temperature at a ramp rate of 0.3–0.6°C/min. The MOCVD growth of InSb/InAs QDs was performed in a standard horizontal reactor under atmospheric pressure at temperatures of 420–440°C and at the rate of TMIn supply to the reactor of 5.6–8 μmol/min. The details of the QD growth were reported elsewhere [1–5].

The samples with arrays of InSb QDs on the InAs surface were prepared for the study by TEM using the conventional procedure of preliminary mechanical thinning followed by final ion beam milling (4 keV Ar^+ ions). The TEM study was carried out in a JEOL JEM 2100F microscope at an accelerating voltage of 200 kV in the diffraction and high-resolution modes. Both cross-section and plan-view TEM specimen configurations were used.

3. TEM results

TEM observation showed the QD ensemble to have bimodal size distribution [2]. There were small QDs of the typical size of 5 nm in height and 18–22 nm in diameter as well as big QDs of 9–10 nm in height and 38–50 nm in diameter. The majority of QDs looked lens-like in the cross-section TEM image. Also there were some QDs having a truncated spherical shape.

Under two-beam diffraction conditions the regular moiré fringes appeared in plan-view TEM images of small InSb QDs. Such fringes were shown to occur [6-7] due to fine difference in atomic plane distance in the InAs matrix and elastically relaxed free-standing QD. An example of the fringed image obtained from small InSb QD under two-beam diffraction conditions is shown in figure 1 (a). The fringes always run perpendicular to the diffraction vector, and when the diffraction vector rotates to orthogonal direction the fringes lay-out also becomes orthogonal.

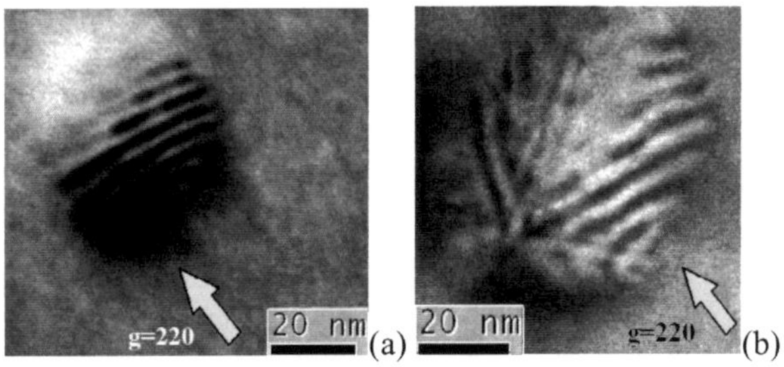

Figure 1. Plan-view TEM images with the moiré pattern of small (a) and big (b) InSb/InAs QD

As distinct from small QDs the big ones were discovered to exhibit whether the complicated moiré pattern (figure 1 (b)) or the specific "tick"-like contrast features. Additionally, the fringe pattern from big QD did not rotate accordingly to the rotation of the diffraction vector ***g*** but showed a significant change as can be seen in figure 2.

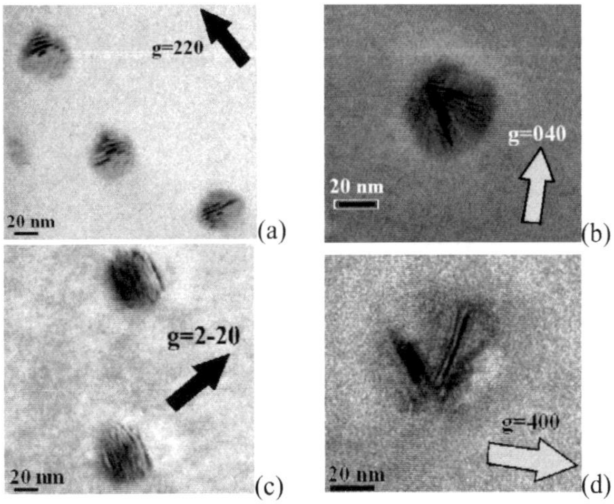

Figure 2. TEM images of big InSb/InAs QDs taken under various diffraction conditions

The study of big QDs in high-resolution mode using the cross-section specimen revealed the presence of dislocations near the QD-substrate interface. Such a dislocation is seen in the cross-sectional image of big QD presented in figure 3.

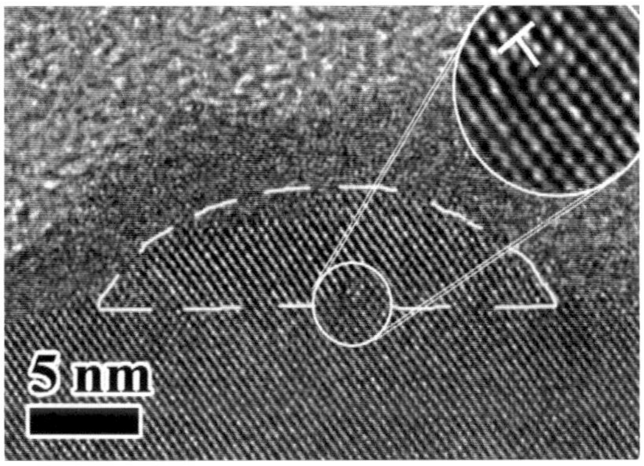

Figure 3. HREM image of InSb/InAs QD with a misfit dislocation

4. Modeling

In order to understand the observed features of TEM/HREM image a theoretical structure modeling and subsequent simulation of image contrast is usually required. To interpret the complicated moiré pattern we have calculated diffraction contrast from QD containing a misfit defect.

A partial Frank dislocation (FD) often occurring in incoherent QDs in the well studied InAs/GaAs system [8] was selected to represent a misfit defect in our model. The dislocation with the Burgers vector b=⅓[111] was placed at the QD-InAs substrate interface along [110] direction and displacement fields inside and outside QD were calculated relative to unstrained part of InAs substrate using the finite element method.

We examined QD of a cylindrical, a lens-like and a truncated spherical shape. The last two cases – a lens-like shape and a truncated sphere – were observed in TEM.

The results of the displacement field calculation for the developed model have demonstrated that the QD shape has insignificant influence on the magnitude and distribution of the displacement field.

The insertion of FD into QD of each shape essentially changes the distribution of the local displacement field and reduces strain at the edges of QD almost by 30%. As an example, the calculated displacement field for cylindrical QD is exhibited in figure 4. One can compare the position of equal displacement lines for QD without FD in figure 4 (a) and for QD with FD in figure 4 (b).

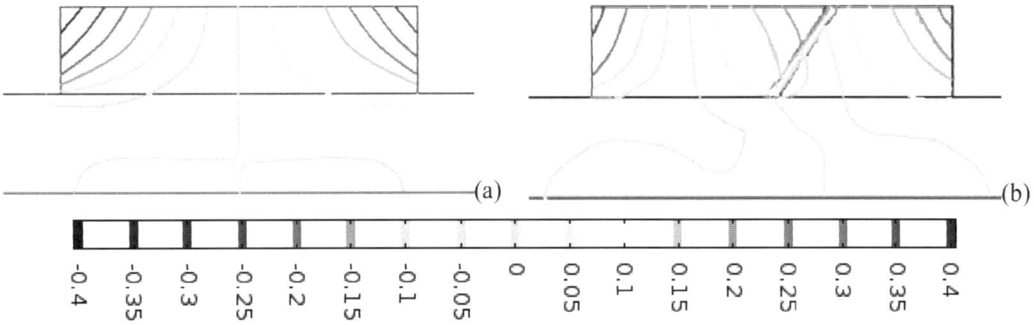

Figure 4. Maps of radial displacements in QD without FD (a) and with FD (b)

Calculated displacements were then used in simulation of diffraction contrast TEM images. The Howie-Whelan dynamic approach [9–10] was applied to simulate the plan-view image of QD containing FD. The approach is commonly used for the solution of electron diffraction problem in the crystals containing defects and is based on two differential equations for the amplitudes of incident (Φ_o) and diffracted (Φ_g) electron waves:

$$\frac{\mathrm{d}\Phi_o(z)}{\mathrm{d}z} = \frac{\pi i}{\xi_g}\Phi_g(z)\exp(2\pi i\vec{g}\cdot\vec{u})$$
$$\frac{\mathrm{d}\Phi_g(z)}{\mathrm{d}z} = \frac{\pi i}{\xi_g}\Phi_o(z)\exp(-2\pi i\vec{g}\cdot\vec{u}) + 2\pi i\vec{s}\,\Phi_g(z) \quad (1)$$

where $\vec{g}$ is the diffraction vector, ξ_g is the extinction distance, $\vec{s}$ is the parameter of deviation (in our calculations $\vec{s} = 0$).

5. Results and discussion

The results of TEM image simulation showed that the shape of QD had insignificant influence on the final diffraction contrast.

Behavior of diffraction contrast in the simulated TEM image depending on the operating diffraction vector and a lattice mismatch value is presented in figure 5.

As can be seen, the insertion of FD into QD leads to various changes in the moiré pattern. It causes the appearance of additional moiré fringes in the diffraction contrast image with the diffraction vector of $g = 220$. The introduction of FD also breaks the symmetry of the image with the diffraction vector of $g = 400$ or 0–40.

The increase in a lattice mismatch results in reduction of the moiré fringes period and, therefore, leads to the increase in the number of observable fringes.

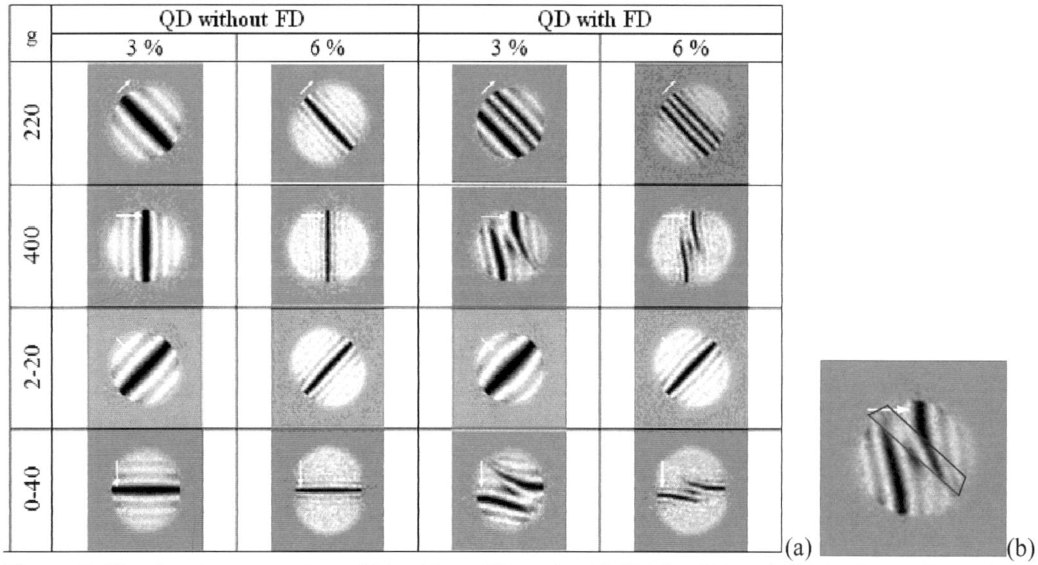

Figure 5. Simulated contrast from QD without FD and with FD for 3% and 6% lattices mismatch (a) and the position of stacking fault (it is shown by the counter) associated with FD in QD (b).

Comparison of simulated and experimental images exemplified in figure 6 shows that the observed period of the moiré pattern from small QD under 220 reflection (figure 6 (e–f)) is very close to that in the simulated image of QD with a 6% lattices mismatch (figure 6 (l–m)). It means that QD consists of pure InSb and does not contain arsenic (or a quite low As concentration).

The characteristic "tick"-like contrast observed in the experimental TEM image taken from big QD under the operating diffraction vector 400 occurs in simulated images of QD containing FD at the interface (figure 6 (a–d) and figure 6 (g–k)).

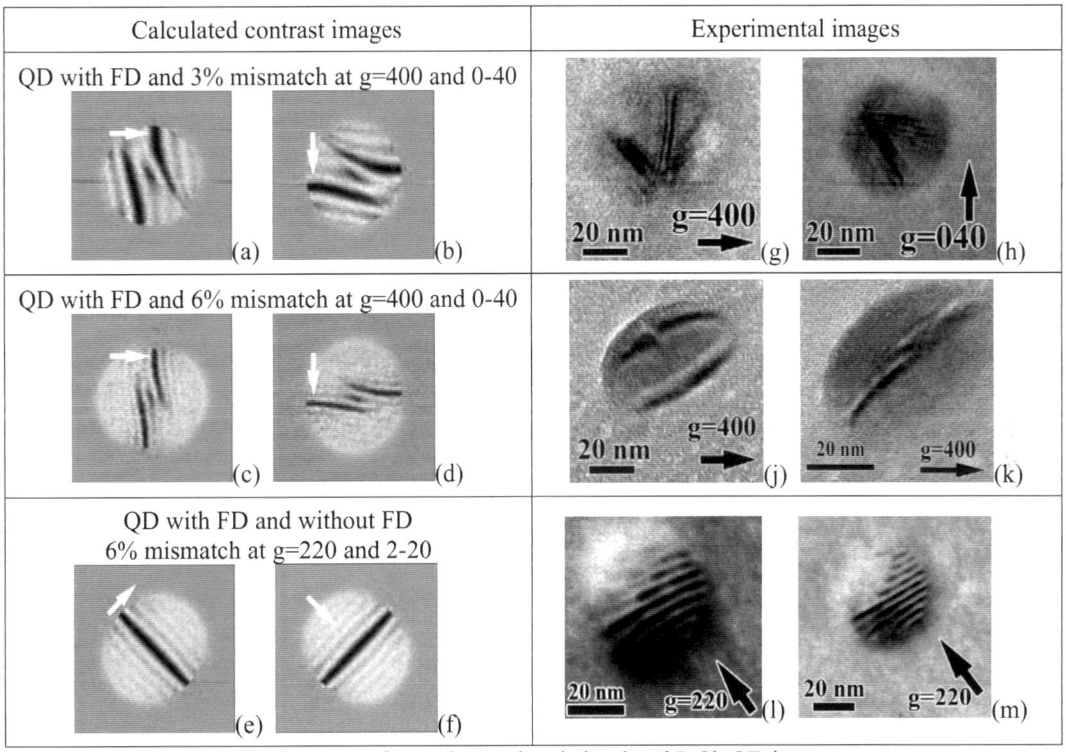

Figure 6. Comparison of experimental and simulated InSb QD image contrast.

6. Conclusion

Some structural features of InSb QDs free-standing on InAs substrate were investigated by TEM supplemented with image simulation. The specific contrast features were observed in the plan-view diffraction images of big QDs (9–10 nm in height and 38–50 nm in diameter). To understand the origin of such distortions in the TEM image the model of InSb QD on InAs substrate was developed and used for calculations of the displacement field and the subsequent diffraction image simulation of InSb QD for the first time.

The shape of QD was established to influence insignificantly the magnitude of radial displacements. The insertion of a misfit defect (a partial Frank dislocation) into QD reduces the strain at the edges of QD almost by 30%.

The simulated images of QD correlate well with the specific "tick"-like contrast in the experimental electron micrographs in the case of a Frank dislocation inserted into the QD-substrate interface.

References

[1] Mikhailova M, Stoyanov N, Andreev I, Zhurtanov B, Kizhaev S, Kunitsyna E, Salikhov K and Yakovlev Yu 2007 *Proc. SPIE* **6585** 658526

[2] Moiseev K D, Parkhomenko Y A, Gushchina E V, Ankudinov A V, Mikhailova M P, Bert N A and Yakovlev Y P 2009 *Semiconductors* **43** 1102–09

[3] Moiseev K D, Parkhomenko Y A, Ankudinov A V, Gushchina E V, Mikhailova M P, Titkov A N and Yakovlev Y P 2007 *Tech. Phys. Lett.* **33** 295–98
[4] Romanov V V, Dement`ev P A and Moiseev K D 2013 *Semiconductors* **47** 443–48
[5] Bert N A, Nevedomsky V N, Dement'ev P A and Moiseev K D 2013 *Appl. Surf. Sci.* **267** 77-80
[6] Androussi Y, Benabbas T and Lefebvre A 2002 *Ultramicroscopy* **93** 161
[7] Bert N A, Freidin A B, Kolesnikova A L, Korolev I K and Romanov A E 2010 *Phys. Status Solidi A-Appl. Mat.* **207** 2323–26
[8] Chen Y, Lin X W, Liliental Z, Washburn J, Klem J F and Tsao J Y 1996 *Appl. Phys. Lett* **68** 111
[9] Hirsch P, Howie A, Nicholson R B, Pashley D W and Whelan M J 1977 *Electron Microscopy of Thin Crystals* (New York: Robert E. Krieger Publishing)
[10] David B Williams and C Barry Carter 2009 *Transmission Electron Microscopy A Textbook for Materials Science Part 2* (New York: Plenum Press)

Terahertz reflection and emission associated with nonequilibrium surface plasmon polaritons in *n*-GaN

G A Melentyev[1], V A Shalygin[1], M D Moldavskaya[1], V Yu Panevin[1], L E Vorobjev[1], D A Firsov[1], H Nykänen[2], L Riuttanen[2], O Svensk[2] and S Suihkonen[2]

[1] Department of Physics of Semiconductors & Nanoelectronics, St.Petersburg Polytechnic University, Polytechnicheskaya 29, St.Petersburg 195251, Russia
[2] School of Electrical Engineering, Aalto University, Otakaari 1, Espoo 02150, Finland

E-mail: gamelen@spbstu.ru

Abstract. Surface plasmon polaritons are investigated in heavily doped *n*-GaN epitaxial layers. The grating etched on the surface of the epitaxial layer is used to convert photons into the surface plasmon polaritons and vice versa. The spectral study of reflection demonstrates the possibility of nonequilibrium surface plasmon polaritons excitation due to terahertz radiation scattering on the grating. Terahertz electroluminescence is investigated under lateral electric field. The luminescence spectrum demonstrates a significant contribution of nonequilibrium surface plasmon polariton scattering to terahertz radiation emission.

1. Introduction

Plasmonics provides ample opportunities for a variety of highly significant applications in modern science and technology. Signal processing components, emitters and detectors for modern photonics, subwavelength-sized elements for chemical and biological sensing with extremely high sensitivity (up to single molecular sensing), effective photovoltaic devices and solar cells are developed on the base of plasmonics [1–6]. The distinctive objects in plasmonics are surface plasmon polaritons (SPP). They represent electromagnetic excitations propagating at the interface between a conductor and a dielectric, evanescently confined in the perpendicular direction. These electromagnetic surface waves arise via the coupling of the electromagnetic fields to oscillations of the free carrier density in the conductor. Coupling of photons to charge carriers at conductor interfaces allows subdiffraction-limit localization of electromagnetic radiation and strong field enhancement. This high field localization occurs as long as the fields oscillate at frequencies close to the intrinsic plasma frequency of the conductor [1]. Many plasmonic phenomena are studied on metals such as gold, silver, or aluminium, and in such a case resonant plasma oscillations are in the visible spectral range. The modern trend in the field of plasmonics is mid infrared (mid-IR) and terahertz (THz) plasmonics which deals with heavily doped semiconductors instead of metals [5, 7]. In semiconductors the frequency of SPP can be tuned down to THz range by doping level.

One of the first experimental studies of SPP in mid-IR range was performed on *n*-InSb samples with a grating at their surface [8]. The sharp dips associated with SPP excitation were experimentally observed in the reflectivity spectra. The terahertz emission mediated by surface plasmon polaritons in doped semiconductors with regular surface grating was theoretically predicted in [9]. To the best of our knowledge, this phenomenon was not experimentally observed so far. On the other hand, THz electroluminescence was investigated in degenerate *n*-InN epitaxial layers with a *random* grating formed by topographical defects [10]. It was shown that the observed THz radiation emission can be attributed to SPP scattering on the random grating. A similar mechanism of THz electroluminescence is anticipated in heavily doped semiconductors with a *regular* surface grating.

The goal of the present paper is to study the optical phenomena associated with nonequilibrium SPP in heavily doped GaN-based microstructures. Reflection investigation demonstrates excitation of SPP under THz irradiation of samples with a regular surface grating. Investigation of THz electroluminescence demonstrates excitation of SPP under lateral electric field which is accompanied by an impressive increase of radiation intensity in the grating samples.

2. Experimental details

Gallium nitride epitaxial layers were grown on sapphire substrate by MOVPE. The thickness d of epilayers was 6.2 μm. We studied samples doped by Si with the electron concentration of $3.6\cdot10^{19}$ cm^{-3} and the mobility of 122 cm^2/V·s (at room temperature). It is known that such heavily doped GaN layers demonstrate formation of an impurity band and its overlapping with the conduction band [11]. That is why the free electron concentration is constant in the range from helium to room temperature. A grating with a period of Λ = 86 μm was etched on the surface of epitaxial layer. The etch depth was h = 4.5 μm. The ratio of groove width a to grating period Λ was 1 : 2. The design of the sample with a regular surface grating is shown in figure 1. The reference sample without a grating on GaN/air interface was additionally studied.

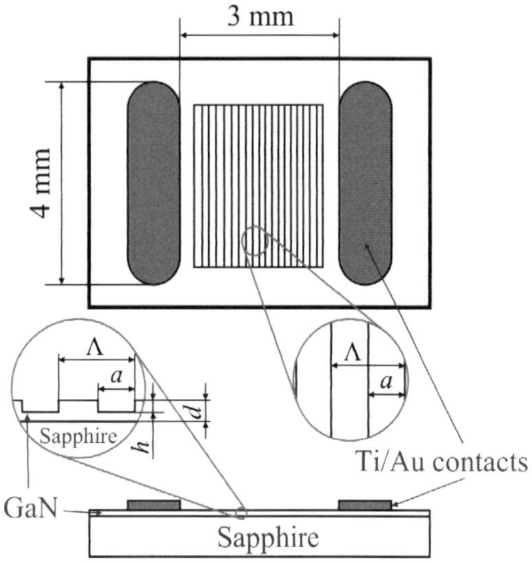

Figure 1. Design of the sample with a regular surface grating. The period of the grating Λ is 86 μm, the groove width a is 43 μm, the etch depth h is 4.5 μm, the thickness of epitaxial layer d is 6.2 μm.

The reflectivity spectra were investigated at room temperature using a Fourier spectrometer operating in a rapid-scan mode and a pyroelectric detector. The globar and the mercury lamp were used as sources of THz radiation. The incidence angle θ was about 11°, and the incidence plane was perpendicular to the grooves of the grating. The spectra were recorded for p- and s-polarized radiation. The experimental setup for reflection study is shown in figure 2.

Emission of THz radiation from the sample has been observed under applying lateral electric field. Measurements were made in a pulsed regime. The voltage pulses with duration of 2 μs and repetition frequency of about 87 Hz were applied. The spectral studies of the emitted radiation were carried out at the temperature of 9 K using a Fourier spectrometer operating in a step-scan mode. A silicon bolometer was used as a detector. The THz radiation was collected in the direction perpendicular to the sample surface, the angular aperture was $\Delta\theta \approx 16°$.

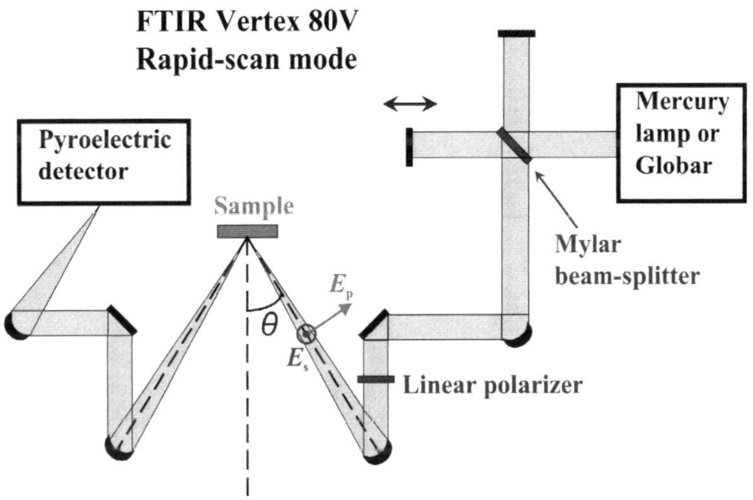

Figure 2. The experimental setup for reflection studies.

3. Results and discussion

Let us consider the surface plasmon polaritons at plane interface between two semi-infinite mediums: heavily doped *n*-GaN (conductor) and vacuum (dielectric). The dispersion law for these SPP can be described by the following expression [1]:

$$k_{\mathrm{SPP}}(\omega) = \frac{\omega}{c}\mathrm{Re}\left(\sqrt{\frac{\varepsilon_1(\omega)\varepsilon_2(\omega)}{\varepsilon_1(\omega)+\varepsilon_2(\omega)}}\right), \tag{1}$$

where k_{SPP} is wave vector of SPP, ω is angular frequency, $\varepsilon_1(\omega)$ is permittivity of *n*-GaN, $\varepsilon_2 = 1$ is permittivity of vacuum. The depth of SPP field decay (in GaN in the direction perpendicular to the interface) estimated in accordance with [1] does not exceed 2 μm (at the frequency more than 1 THz). This value of decay depth is less than the thickness of the studied epitaxial layer. The latter is 3.95 μm for the sample with a regular grating and 6.2 μm for the reference sample. Due to this the relationship expression (1) can be used for the description of SPP in the epitaxial layers considered.

In our experiments we deal with a "non-radiative" mode of surface plasmon polaritons for which k_{SPP} is larger than the wave vector of THz photon at the same frequency. Nevertheless, SPP can produce THz photons under scattering on the regular grating. To achieve this, two conditions must be satisfied. Firstly, the energy of a THz photon must be equal to the energy of SPP (the energy conservation law):

$$\hbar\omega_{\mathrm{THz}} = \hbar\omega_{\mathrm{SPP}} \equiv \hbar\omega, \tag{2}$$

where ω denotes the joint angular frequency of the both quasiparticles. Secondly, phase-matching condition for SPP and photons must be fulfilled [1]:

$$k_{\mathrm{SPP}} = k_0 \sin\theta + m\frac{2\pi}{\Lambda},\ m = \pm1, \pm2, \pm3, \ldots, \tag{3}$$

where $k_0 = \omega/c$ is wave vector of a THz photon, θ is angle of radiation incidence, $2\pi/\Lambda$ is the reciprocal vector of the grating, Λ is grating period. Relation (3) is equivalent to the momentum conservation law.

If we consider backward transformation of the THz photon into the surface plasmon polariton, again the both conditions (2) and (3) must be satisfied. This process can take place under THz radiation reflection from the sample with a grating. The reflectivity decrease is expected at certain THz photon energies.

Let us first analyze the experimental results on reflection. We have measured the reflectivity spectra outside the GaN reststrahlen band at the photon energies $\hbar\omega < \hbar\omega_{TO}$, where $\hbar\omega_{TO} = 69.3$ meV [12] is the energy of the transverse optical phonon in wurtzite GaN (E_1 symmetry). The experimental spectra for two linear polarizations are presented in figure 3. The reflectivity spectrum for *p*-polarization demonstrates a set of dips, whereas the *s*-polarization spectrum does not demonstrate them. It should be noted that the SPP is the *TM*-type electromagnetic wave, so in our experimental conditions (see figure 2, the incidence plane is perpendicular to the grating grooves) the surface plasmon polaritons can be excited only by *p*-polarized radiation. In figure 3, the arrows denote the positions where the energy conservation law (2) and the phase-matching condition (3) at the angle of 11° for an incident photon and SPP are realized. One can see a good correlation between theoretically expected positions of the reflectivity dips and experimentally observed ones. Thus the polarization peculiarities of reflection prove the excitation of nonequilibrium surface plasmon polaritons by the incident THz radiation.

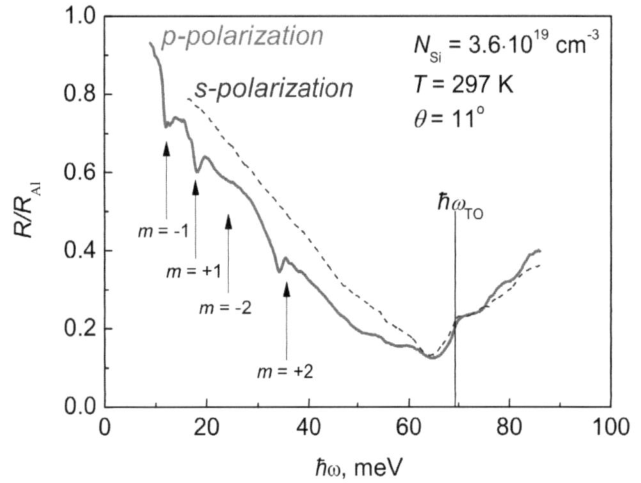

Figure 3. Reflectivity spectra of the epitaxial layer GaN with a grating normalized to an aluminum mirror. Solid (dashed) line demonstrates the experimental results for *p*-polarization (*s*-polarization). The arrows denote the positions where the energy conservation law (2) and the phase-matching condition (3) at the angle of 11° for an incident photon and SPP are satisfied.

The emission of terahertz radiation was investigated in GaN epilayers under applying lateral electric field. The microscopic origin of the THz emission is associated with field-induced deviation of the electron and SPP ensembles from the equilibrium distribution. THz emission associated with two dimensional (2D) plasmons was experimentally observed in Si MOSFETs and AlGaAs/GaAs hetoreinterfaces [13–15]. We investigated the spectral radiation density for *n*-GaN epitaxial layer with a grating $D_\omega(h)$ and the reference sample without a grating $D_\omega(0)$. Samples had the same concentrations of free electrons. In the samples without a grating the observed terahertz radiation is related to the blackbody-like radiation emission from hot electrons. We have previously studied this phenomenon in GaN-based nanostructures [16]. The spectral density of THz emission from the grating sample $D_\omega(h)$ normalized to the sample without a grating under dissipated electric power 1500 W is displayed in figure 4. This dependence demonstrates sequence of peaks. The arrows denote the positions where the energy conservation law (2) and the phase-matching condition (3) at the angle $\theta = 0°$ for an emitted THz photon and SPP are fulfilled. The arrows are close to the observed THz

Journal of Physics: Conference Series **586** (2015) 012005 doi:10.1088/1742-6596/586/1/012005

emission peaks. The dispersion dependence for surface plasmon polaritons calculated according to (1) is demonstrated in figure 5. Gray fill indicates energy ranges (and corresponding wave numbers) of the SPP where conditions (2) and (3) are satisfied for the angular range of $-8° \leq \theta \leq +8°$. In these energy ranges THz photon emission due to nonequilibrium SPP scattering on the grating can be observed. As it is seen from figure 5, an increase of the factor m in (3) is accompanied by an extension of energy range where THz photons can be emitted. The experimental data on the ratio $D_\omega(h)/D_\omega(0)$ shown in figure 4 are consistent with the theoretical calculation displayed in figure 5. In particular, the experimentally observed radiation linewidths are less or equal to the calculated values, with increase of m the radiation linewidths increases. All the above-mentioned results confirm that THz emission is associated with the scattering of nonequilibrium surface plasmon polaritons on a grating.

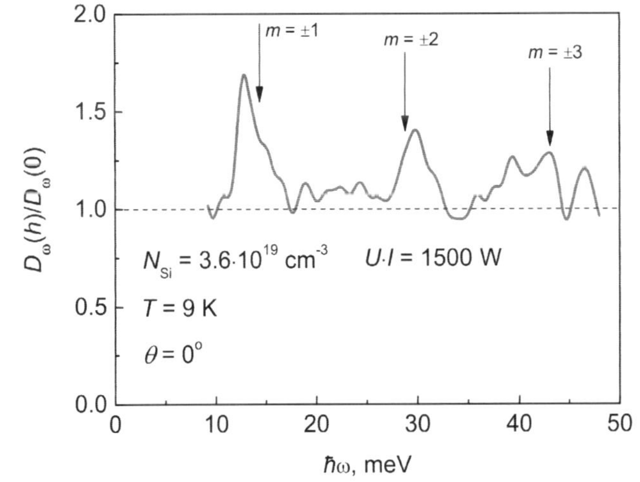

Figure 4. The spectral radiation density ratio for the sample with a grating $D_\omega(h)$ and the reference sample without a grating $D_\omega(0)$ at the same dissipated electric power 1500 W. The arrows denote the positions where the energy conservation law (2) and the phase-matching condition (3) at the angle $\theta = 0°$ for an emitted photon and SPP are satisfied.

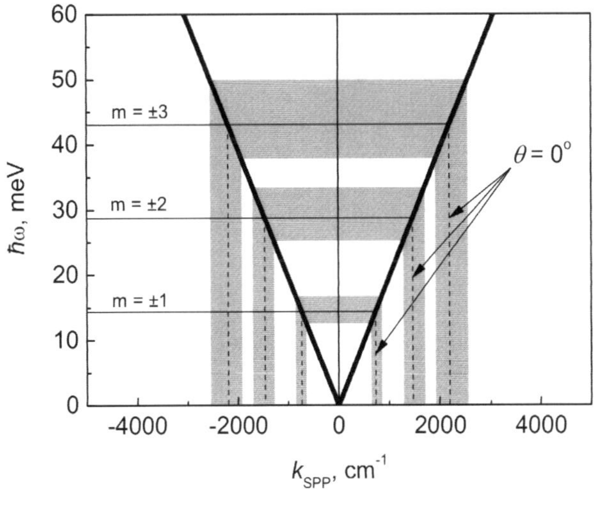

Figure 5. Dispersion dependence for surface plasmon polaritons. Gray fill indicates energy ranges (wave numbers) of the SPP where the energy conservation law (2) and the phase-matching condition (3) for an emitted photon and SPP are satisfied in the case of angular aperture $\Delta\theta = 16°$ (angle $\theta = -8° \ldots +8°$).

We obtained the good agreement between theoretical evaluations and experimental data. It proves that the expression (1) describing surface plasmon polaritons at plane interface between semi-infinite conductor and vacuum can be used adequately for the considered n-GaN/vacuum interface (thickness of a few micrometers, a regular grating on the epilayer surface).

4. Conclusions

We have investigated the reflection and electroluminescence spectra of heavily doped GaN-based microstructures. The samples with a regular grating on a surface of the microstructure, as well as with the flat surface GaN/vacuum have been studied.

The investigation of optical reflection demonstrated the possibility of nonequilibrium SPP excitation under THz radiation scattering on a surface grating. The study of electroluminescence spectra demonstrated a significant contribution of nonequilibrium SPP scattering on the grating to THz radiation emission. The good agreement between theoretical evaluations and experimental results has shown that the behavior of surface plasmon polaritons on the interface GaN/vacuum in investigated epitaxial layers can be successfully described by the expressions for the semi-infinite mediums. Our studies can be applied for the development of portable sources of THz radiation operating under electric pumping.

Acknowledgements

The authors are grateful to A.N. Sofronov for the help in the experiments. This work was supported by the Russian Foundation for Basic Research (grant 14-02-90444), Russian Federation President grant for young scientists and the Ministry of Education and Science of the Russian Federation.

References

[1] Maier S A 2007 *Plasmonics: Fundamentals and Applications* (New York: Springer)
[2] Berrier A, Ulbricht R, Bonn M and Rivas J G 2010 *Opt. Express* **18** 23226
[3] El Fatimy A, Dyakonova N, Meziani Y, Otsuji T, Knap W, Vandenbrouk S, Madjour K, Théron D, Gaquiere C, Poisson M A, Delage S, Prystawko P and Skierbiszewski C 2010 *J. Appl. Phys.* **107** 024504
[4] Mayer K M and Hafner J H 2011 *Chem. Rev.* **111** 3828
[5] Ginn J C, Jarecki R L Jr, Shaner E A and Davids P S 2011 *J. Appl. Phys.* **110** 043110
[6] Krenn J 2012 *Nature Photon.* **6** 714
[7] Taliercio T, Ntsame Guilengui V, Cerutti L, Rodriguez J B and Tournié E 2013 *Proc. SPIE* **8807** 880702
[8] Marschall N, Fischer B and Queisser H J 1971 *Phys. Rev. Lett.* **27** 95
[9] Kong B D, Sokolov V N, Kim K W and Trew R J 2008 *J. Appl. Phys.* **103** 056101
[10] Shubina T V, Gippius N A, Shalygin V A, Andrianov A V and Ivanov S V 2011 *Phys. Rev.* B **83** 165312
[11] Agekyan V F, Vorob'ev L E, Melentyev G A, Nykänen H, Serov A Yu, Suihkonen S, Filosofov N G and Shalygin V A 2013 *Phys. Solid State* **55** 296
[12] Karch K, Wagner J-M and Bechstedt F 1998 *Phys. Rev.* B **57** 7043
[13] Höpfel R A, Vass E and Gornik E 1982 *Phys. Rev. Lett.* **49** 1667
[14] Okisu N, Sambe Y and Kobayashi T 1986 *Appl. Phys. Lett.* **48** 776
[15] Hirakawa K, Yamanaka K, Grayson M and Tsui D C 1995 *Appl. Phys. Lett.* **67** 2326
[16] Shalygin V A, Vorobjev L E, Firsov D A, Sofronov A N, Melentyev G A, Lundin W V, Nikolaev A E, Sakharov A V and Tsatsulnikov A F 2011 *J. Appl. Phys.* **109** 073108

Analysis of electron energy spectrum in type II core/shell quantum dots

A A Karmanov, I A Pronin, N D Yakushova, S E Igoshina and I A Averin

Penza State University, 40 Krasnaya Street, Penza 440026, Russia

E-mail: sigoshina@mail.ru

Abstract. We investigated the peculiarities of the electron energy spectrum in type II core/shell quantum dots. It is shown that the presence of the shell with energy bands shifted relative to the ones in the core gives additional possibilities to control energy level positions.

1. Principal types of core/shell quantum dots

One of the main objects of low-dimension semiconductor heterostructure physics is so-called quasi-zero-dimensional systems or quantum dots (QDs). There have been a lot of experimental [1, 2] and theoretical [3, 4] researches of QDs with various geometrical shapes. Recent nanotechnology trends are related to the generation of core/shell QDs [5, 6]. Like in the classical heterotransition theory, two main core/shell QD types can be pointed out.

Energy gaps in semiconductors forming type II QDs (figure 1) are comparable but the band edges are shifted one relative to another. It is energetically favorable for photo-excited charge carriers to stay within different regions of such a QD, e.g. for an electron to stay inside the shell, and for a hole to stay inside the core.

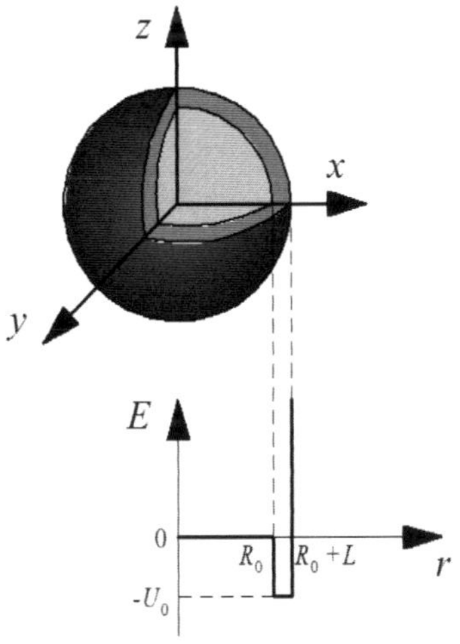

Figure 1. Scheme of type II core/shell quantum dot and the electron potential energy vs distance.

Such quantum dots show the spatial separation of charge carriers and thus they are promising for the systems with a long lifetime of electron-hole pairs due to the reduced overlapping of the electrons and holes wave functions. Other effects in type II QDs are an effective energy gap narrowing and a luminescence maxima shift to the red spectral range.

The shell of a type I QD consists of a wide band semiconductor, while the core is a narrow band one. This passivates surface states and localizes electron-hole pairs inside the core. This approach is used to increase the QD luminescence effectiveness.

The aim of this study is to develop a theory of the energy spectrum in type-II core/shell quantum dots.

2. Electron energy spectrum calculations in Type II core/shell quantum dot

The problem under study is being solved using the rigid wall model. The stationary Schrödinger equation in the effective mass approximation is

$$-\frac{\hbar^2}{2m(r)}\nabla^2\psi + U(r)\psi = E\psi . \tag{1}$$

The potential $U(r)$ and the electron effective mass $m(r)$ inside the QD are expressed in the following way:

$$U(r) = \begin{cases} 0 & r \le R_0 \\ -U_0 & R_0 < r \le R_0 + L \\ \infty & r > R_0 + L \end{cases} , \tag{2}$$

$$m(r) = \begin{cases} m_1^* & r \le R_0 \\ m_2^* & R_0 < r \le R_0 + L \end{cases} , \tag{3}$$

where U_0 is the potential barrier height, m_1^*, m_2^* are the effective masses of electrons inside the core and the shell, R_0 is the core radius, and L is the shell thickness.

Using the boundary condition that the wave function vanishes on the QD boundary the solution of the equation (1) determining the energy spectrum of the electrons with energies of $U_0 < E_2 < 0$ can be obtained in the form of:

$$E_2 = \frac{\hbar^2 \xi_{nl}^2}{2m_2^*(R_0 + L)^2} - U_0 , \tag{4}$$

where ξ_{nl} is the n-th root of the Bessel function of the l-th kind.

The equality and continuity of the wave functions at the QD core/shell border defines the energy spectrum of electrons with energies of $E_l > 0$:

$$m_2 \left.\frac{dj_l(k_1 r)}{dr}\right|_{r=R_0} = m_1 \frac{j_l(k_1 R_0)}{j_l(k_2 R_0)} \left.\frac{dj_l(k_2 r)}{dr}\right|_{r=R_0} . \tag{5}$$

Figure 2 shows the results of numerical analysis of expressions (4) and (5) for electrons in the type II core/shell CdTe/CdSe quantum dots with the following parameters: the effective masses of electrons are $m_{CdTe} = 0.14 m_0$ and $m_{CdSe} = 0.13\ m_0$, where m_0 is the free-electron mass; $U_0 = 0.60$ eV; $l = 1$. The analysis shows that the number of discrete levels with $U_0 < E_2 < 0$ is finite and, moreover, the increase in core thickness from 2 to 3 nm at a fixed QD radius leads to an appearance of extra energy levels. When R_0 values are low, the probability of the electron location in the QD core is low as well, that leads to energy oscillations (figure 2 (a)). At higher core radius electrons occupy stable positions, oscillations are terminated, and charge carrier energies decrease significantly (figure 2 (b)). Furthermore, it can be seen that as the size of the QD core increases, the energy levels converge, and the quantum size effects at room temperature for $R_0 > 10$ nm become unobservable.

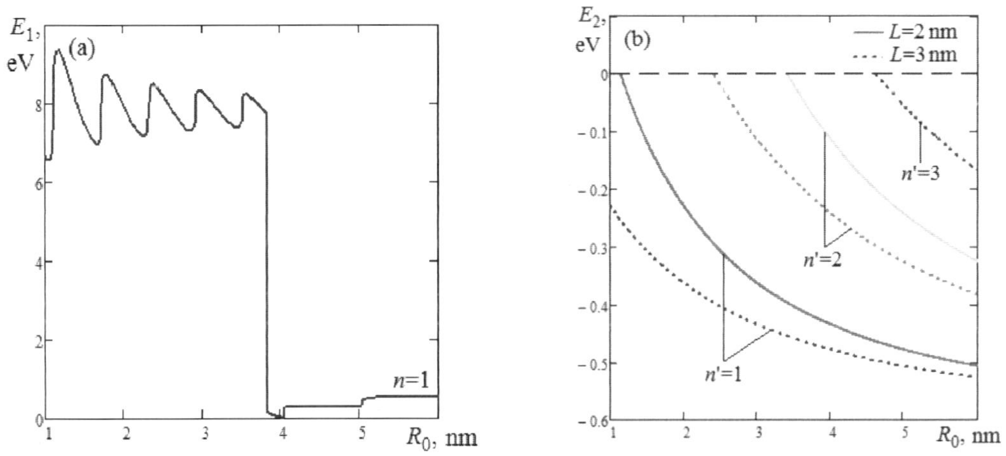

Figure 2. Energy spectrum of the electron as a function of the radius of the type II core/shell quantum dot: (a) $E_1 > 0$; (b) $-U_0 < E_2 < 0$.

The results obtained show a complicated dependence of the electron energy spectrum on the QD radius. Due to the presence of the shell with the energy bands shifted relative to the ones in the core, there appear additional possibilities to control the positions of the energy levels.

3. Conclusions

In this study we suggest a theory of the energy spectrum in type-II core/shell quantum dots, according to which there are two groups of discrete energy levels corresponding to energies of $E_1 > 0$ and $-U_0 < E_2 < 0$. As an example, the numerical calculations for electrons in the CdTe/CdSe QD were made and their results were analyzed. The developed model can be used to obtain quantum dots with electrophysical and optical properties needed for a new generation of optical and nanoelectronic devices.

This work is financially supported by the Russian Ministry of Education, the state task number 2014/151 (project 117).

References

[1] Kochereshko V P, Kats V N, Platonov A V, Suris R A., Cirlin G E, Buravlev A D, Samsonenko Y B, Besombes L, Le Gal C and Mariette H 2012 *Journal of Surface Investigation* **6** 722–725

[2] Tarasov S A, Aleksandrova O A, Maksimov A I, Marayeva E V, Matyushkin L B, Menkovich E A, Moshnikov V A and Musikhin S F 2013 *Izvestia Vysshik Uchebnykh Zavedenii. Elektronika, in Russian* **101(3)** 28–32

[3] Igoshina S E and Karmanov A A 2013 *Quantum Electronics* **43(1)** 76–78

[4] Bolshakov A D, Dubrovskii V G, Yan X, Zhang X and Ren X 2013 *Technical Physics Letters* **39** 1047–1052

[5] Oleinikov V A, Sukhanov A V and Nabiyev I R 2007 *Nanotechnologies in Russia* **2(1)** 160–73

[6] McElroy N, Page R C, Espinbarro-Valazquez D, Lewis E, Haigh S and Binks D J *Thin Solid Films* **560** 65–70

The adsorption of a hydrogen atom on the two types of boron sheets surface

E V Boroznina, N N Borkhoeva and S V Boroznin

Volgograd State University, 100 University Ave., Volgograd 400062, Russia

E-mail: extrajenya@bk.ru

Abstract. The possible connection between the atom of hydrogen and the surface of two-dimensional boron sheets have been investigated. The calculations were carried out within the model of molecular cluster with the use of quantum chemical MNDO scheme. Two types of BS were studied: triangular BS (TBS) and α-sheet of boron atoms (αBS). The hydrogen atoms attacking BS have been simulated by a step-by-step approach for all atom locations. The surface patterns of potential energy for these processes were built. The analysis of curves showed that the active atom of hydrogen is adsorbed on the surface of BS. The adsorption distances (R_{ad}) and the adsorption energy (E_{ad}) were calculated. We have proved that αBS has a greater sorption capacity than hydrogen TBS.

1. Introduction

The search for suitable materials for hydrogen storage has driven attention to different light-weight nanostructures. Among them there are carbon nanostructures including carbon nanotubes (CNTs) which were intensively investigated during the last years. Many layered materials demonstrate their ability to form nanotubular structures. These nanotubes could be potential candidates to store molecular hydrogen because of their large surface area. Among the light-weight elements, besides C, boron would be a good candidate. Nowadays, many research teams and laboratories are engaged in the researches connected with boron and its compounds [1].

In 2004, the single-walled boron nanotubes have been synthesized [2]. Recently, scientists are interested in two-dimensional boron sheets (BS) [3]. Synthesis of this novel two-dimensional material attracts considerable interest because of its unique properties. The two-dimensional boron is expected to be used in many fields of science and technology including hydrogen energetic.

2. The adsorption mechanism of hydrogen atoms on the surface of a boron sheets

The first step of our research was the investigation of the binding opportunity between the atom of hydrogen and the surface of two-dimensional BS and the mechanism of this process. The calculations were carried out within the model of molecular cluster with the use of quantum chemical MNDO scheme [4]. We have studied two types of BS: triangular BS (TBS) and α-sheet of boron atoms (αBS). The examples of an αBS and a triangular BS are shown in figure1.

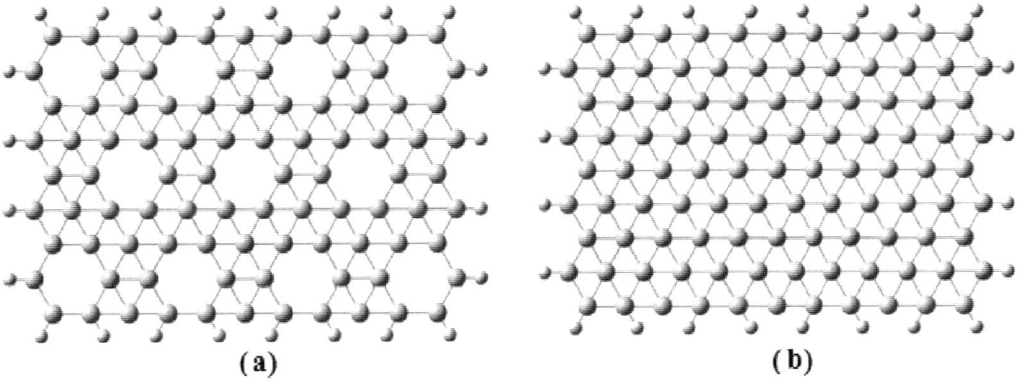

Figure 1. The cluster of pure boron sheets (BS): (a) an α-sheet of boron atoms (αBS); (b) a triangular boron sheet (TBS).

We studied three positions of the adsorbed H atom relative to the αBS surface: (I) over a boron atom in the hexagon node, (II) over a boron atom in the hexagon centre, (III) over the hexagon centre (figure 2) and one position of the adsorbed atom H relative to the TBS surface: (I) over a boron atom (figure 3).

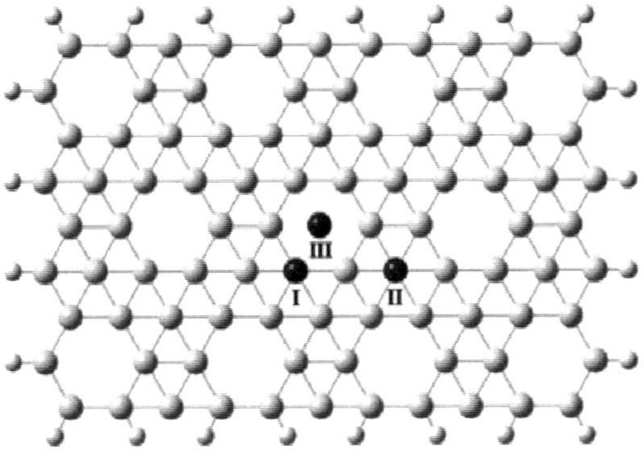

Figure 2. Three positions of adsorbed atoms over the surface of αBS: (I) over a boron atom in the hexagon node, (II) over a boron atom in the hexagon centre, (III) over the hexagon centre.

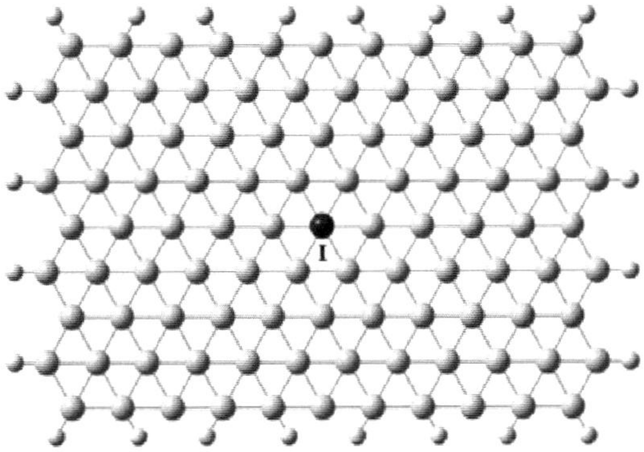

Figure 3. Position of adsorbed atoms over the surface of TBS: (I) over a boron atom.

The process of adsorption for all positions was simulated by step-by-step approach (with a step of 0.1 Å) of the adsorbed atom to a B atom (or a dummy atom) perpendicular to the surface of BS. The calculations have allowed to model profiles of potential energy surface of the given processes (figures 4, 5). The energy curves of these processes show a minimum (excepting the case where an H atom was placed in the position III) which illustrates the fact of a chemical bond formation between H atoms and a B atom on the surface of BS. An H atom is not adsorbed over the centre of a boron hexagon of αBS surface. The basic characteristics of the adsorption process for the hydrogen atom are shown in table 1 and table 2. The analysis of adsorption energy values showed that αBS has a greater sorption capacity with respect to hydrogen than TBS.

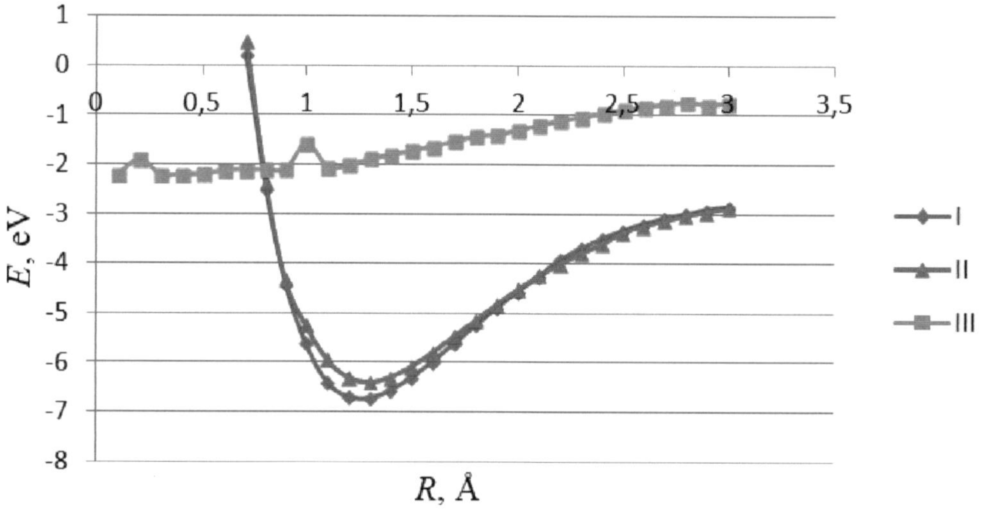

Figure 4. Profiles of surface potential energy of the H atom interaction with the surface of αBS: (I) position I – over a boron atom in the hexagon node, (II) position II – over a boron atom in the hexagon centre, (III) over the centre of a boron hexagon.

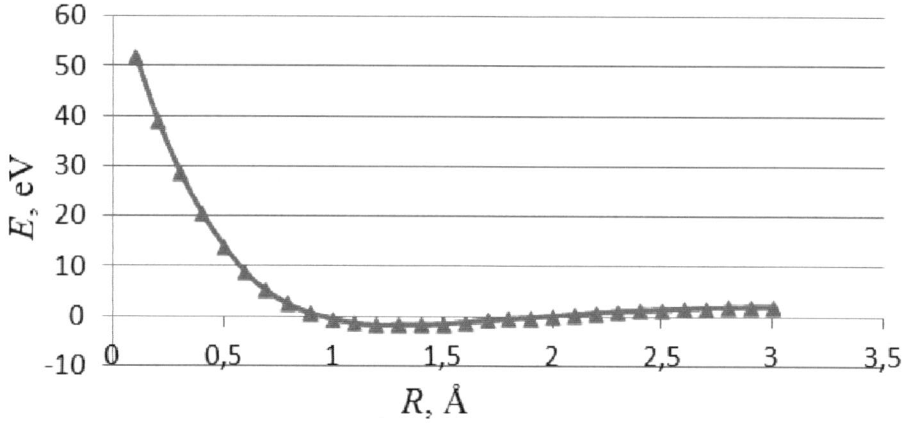

Figure 5. Profile of surface potential energy of the H atom interaction with the surface of TBS: position I – over a boron atom in the hexagon centre.

Table 1 The basic characteristics of the adsorbed H atom over the surface of αBS: (I) over a boron atom in the hexagon node, (II) over a boron atom in the hexagon centre; E_{ad} – the energy of adsorption, eV; R_{ad} – the distance of adsorption, Å; ΔE_g – the width of band gap zone, eV; q_B and q_H – the charges on B and H atoms respectively.

Positions of Hydrogen Atom	E_{ad}, eV	R_{ad}, Å	ΔE_g, eV	q_B	q_H
I	-6,68	1,2	1,98	-0,18	0,03
II	-6,4	1,3	1,91	-0,14	-0,003
αBS			0,09		

Table 2 The basic characteristics of the adsorbed H atom over the surface of TBS: (I) over a boron atom; E_{ad} – the energy of adsorption, eV; R_{ad} – the distance of adsorption, Å; ΔE_g– the width of band gap zone, eV; q_B and q_H – the charges on B and H atoms respectively.

Positions of Hydrogen Atom	E_{ad}, eV	R_{ad}, Å	ΔE_g, eV	q_B	q_H
I	-1,64	1,2	1,53	-0,09	0,01
TBS			1,51		

Also, we investigated the qualitative change of the band-gap of the considered hydrogen composites on the base of BS (figures 6, 7). The width of the energy gap increases in all cases considered. For αBS the band-gap is equal to about 1,9 eV and for TBS the band-gap is equal to about 1,5 eV. However, we should note that calculations were carried out using the molecular model. In this case the value of the band-gap is a little bit higher because the infinity boron sheets are modeled by limited-size clusters. Therefore, we can assume that hydrogen composites based on BS will exhibit semi-conductive properties.

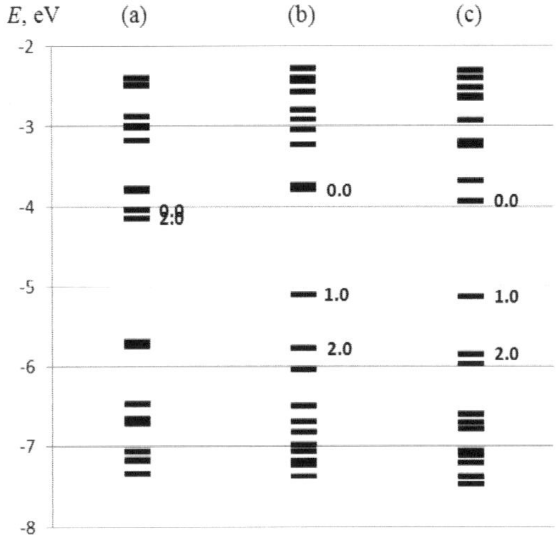

Figure 6. Single electron energy spectra of: (a) pure αBS, (b) αBS with an H atom in the position I, (c) αBC with an H atom in the position II.

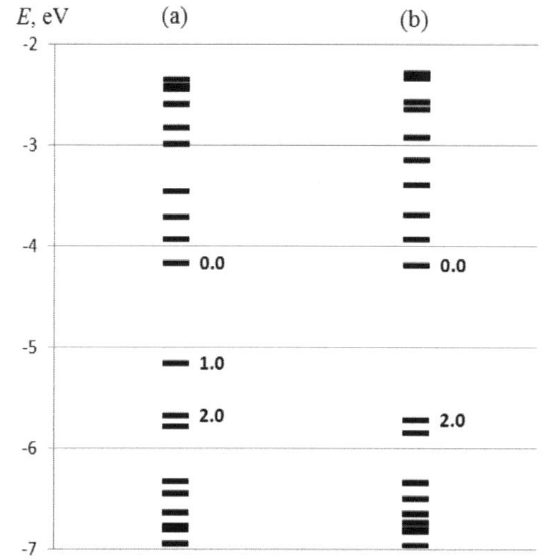

Figure 7. Single electron energy spectra of : (a) pure TBS, (b) TBS with an H atom.

3. The study of regular hydrogenation of boron sheets

The issue of major importance is the question of hydrogenation of surface structures. For graphene this question has been studied in [5, 6]. We consider it necessary to study this phenomenon in boron sheets. We calculated hydrides of two types of BS and studied regular hydrogenation of BS.

To define the most probable hydride structures the calculations for the following two positions of hydrogen atoms relative to a BS surface were carried out: (1) H atoms are located over B atoms, wherein hydrogen atoms are not displaced from each other; (2) the hydrogen atoms are displaced relative to each other for the length of B-B bond (figures 8, 9). The distance of adsorption was taken from the tables 1, 2. The difference of total energies of these positions for both BS is very small and equal to zero within the error limits. So, the analysis of the total energy and charge distribution has showed that the both options of regular hydrogenation for each kind of boron sheets (α and triangular) are available and the new structures are stable. Thus, we can assert that formation of gas-phase hydrogen composites on the basis of boron sheets is possible.

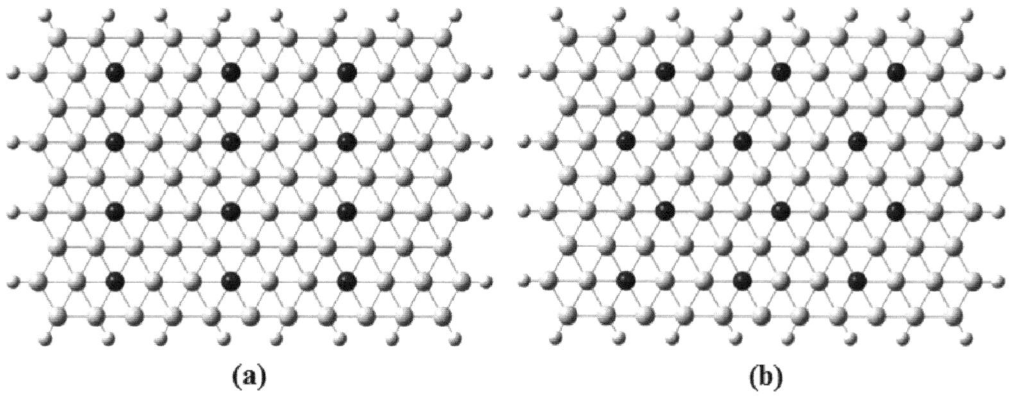

Figure 8. The cluster of TBS with indications of H atom positions on the surface: (a) position 1; (b) position 2.

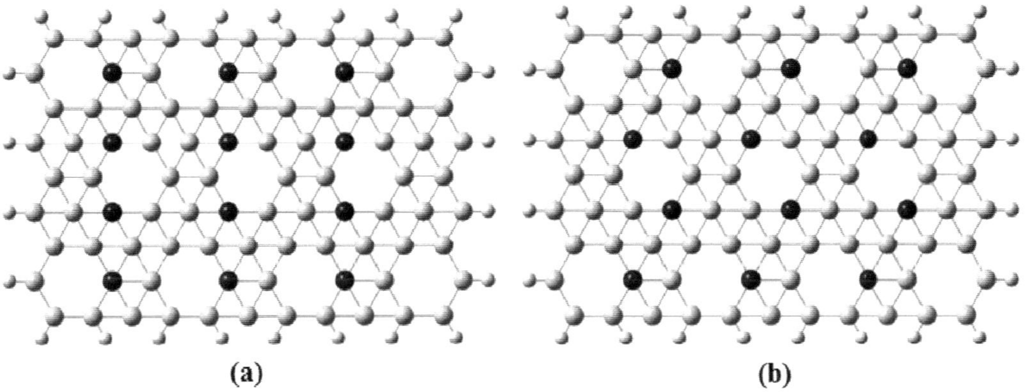

Figure 9. The cluster of αBS with indications of H atom positions on the surface: (a) position 1; (b) position 2.

4. Conclusion

The research of adsorption mechanism for the hydrogen atom on the surface of two-dimensional boron sheets has been investigated. There were three positions of hydrogen adsorption for αBS and one position for TBS. The possibility of adsorption in the position over a B atom has been proofed for both BS. The adsorption energy and the optimal distance of adsorption have been determined.

The process of regular adsorption of hydrogen atoms on the surface of αBS and TBS has been studied. Formation of gas-phase hydrogen composites on the basis of boron sheets is proofed.

References

[1] Cabria I, Lopez M J and Alonso J A 2006 *Nanotechnology* **7** 778785
[2] Ciuparu D 2004 *J. Phys. Chem. B.* **108** 3967 – 9
[3] Liu Y, Penev E S, Yakobson B I 2013 *Angew. Chem* **52** 105
[4] Dewar M J S 1997 *Theoret. Chem. Acta.* **46** 089104

Formation of ordered nanoscale capillary membranes based on anodic alumina

V V Luchinin[1], V A Moshnikov[1,2], E N Muratova[1] and R Sh Samigullin[1]

[1] St. Petersburg Electrotechnical University "LETI", 5 Prof. Popova str., St. Petersburg 197376, Russia
[2] St. Petersburg Polytechnic University, 29 Polytechnicheskaya str., St. Petersburg 195251, Russia

E-mail: sokolovaeknik@yandex.ru

Abstract. Membranes of nanoporous aluminum oxide (alumina) have been obtained using the electrochemical etching technique under varying technological regimes. The basic geometrical parameters including the thickness of the membrane, the average pore size, and the pore size distribution were determined by measuring the ultraviolet-infrared transmittance spectrum. The process of the charged particles transmission through nanoporous alumina membranes with a pore diameter of about 20 nm was studied.

1. Introduction

Currently, the methods of creating nanostructured materials using processes of formation and self-formation are being actively developed [1–4]. One of such materials interesting from a practical point of view is the porous anodic alumina. Membranes based on nanoporous alumina are required in nanotechnology, microbiology, and nuclear physics, because they have unique properties, such as mechanical strength, thermal stability, and chemical resistance.

To form oxide structures based on Si, Al, Ti, and other materials the electrochemical anodization is commonly used. This process is characterized by the availability, as well as the compatibility with the traditional process of micro- and nanoelectronics. However, the electrochemical method of forming micro- and nano-porous alumina most often presented in the literature is realized in thin aluminum films (~ 1 μm). These films are deposited onto the surface of a foreign substrate. The obtained structures are characterized by a large variance in pore size and a nonuniform pore distribution on the surface. The complexity and, in some cases, the impossibility to separate such porous anodic films from substrates limits the possibility of their use as free membranes (masks, matrices) with the required geometrical parameters of the pores.

Thus, formation of mechanically strong membranes based on porous alumina with ordered capillary structure on micro- and nano-scale in a thin (~ 10 μm) aluminum foil is the actual problem.

The main peculiarity of the proposed method is using the thin aluminum foil as an initial substrate for the formation of the autofixed porous alumina membrane. In this case the whole foil is anodized and an additional complex process of the separation of a membrane from a substrate leading to a membrane degradation is not required.

2. Experiment

A well-known mechanism of the ordered honeycomb por-Al_2O_3 formation [1, 5] was taken as a basis in the present work. The main reason for the forces arising between cells is a mechanical stress associated with an increase in alumina volume. The alumina volume expansion can occur only vertically. As a result, the new layer is formed on the pores bottom at the boundary between the aluminum and the oxide layer. Due to this, variations of the voltage and the electrolyte type allow one to increase the aluminum oxide volume by 1.2–1.4 times. The basis for the pore formation effect and subsequent growth of pores is a balance between the electrochemical processes of the local growth and dissolution of aluminum oxide taking place simultaneously on the pore bottom.

In order to obtain a porous anodic alumina membrane [6] the electrochemical anodization of aluminum foil (40 μm) was applied in a potentiostatic mode in electrolytes based on aqueous solutions of phosphoric (H_3PO_4) and sulfuric (H_2SO_4) acids for 10–60 min at temperatures from 0 to 15°C. When using electrolytes with different acidities, the speeds of an anions introduction into the porous layer are different. This effect determines the voltage of the pore formation process and consequently the pore diameter. The anodizing voltage was within the range of 15-30 V (for H_2SO_4) or 130-160 V (for H_3PO_4) depending on the electrolyte type.

The basic structural parameters, such as the pore diameters *a* (see figure 1), the distances between pores *b*, the thickness of the porous layer *c* and the barrier layer *d* of the obtained samples were studied using the scanning electron microscopy (SEM) at low accelerating voltages (10–20 kV) under high vacuum, that allowed us to study the insulating samples without a prior metal deposition.

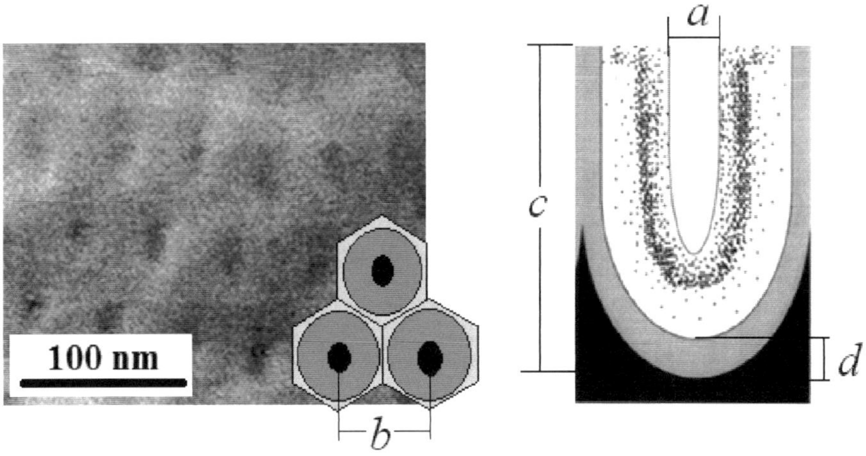

Figure 1. SEM and schematic images of the por-Al_2O_3 membranes.

Figure 2. SEM images of the por-Al_2O_3 membranes obtained in an electrolyte based on solutions of H_3PO_4.

The SEM images of the membranes used in the experiments are shown in figure 2. When the electrolyte based on sulfuric acid is used, the membranes have a thickness of 12-18 μm, the average pore diameter is 20 nm, the barrier layer has a thickness of ≈ 200 nm, and the pore concentration is 350 μm^{-2}. The membranes formed in orthophosphoric acid have the following geometrical parameters: the average pore diameter is 80 nm, the layer thickness is 4–7 μm, the barrier layer thickness is ≈ 200 nm, and the pore concentration is ≈ 50 μm^{-2}. The barrier layers exist in all samples, but they are not etched since high-energy beams were employed in the experiments.

For studying the membranes using the Rutherford backscattering (RBS) and channelling the integral yield of the backscattered ions was measured with the step of 0.1°. The optical transmission measurements were performed using a PE-5400UV spectrophotometer (LLC "Ekohim") in the wavelength range of 190–1000 nm.

3. Results and discussion

The study of the optical transmission spectra have showed that for any membrane the transmission spectra have a similar form [7]: the membrane starts to pass radiation at a certain wavelength. A comparison of the transmission spectra for different membranes shows that the wavelength at which the membrane starts to pass radiation depends on the type of electrolyte used in its production. Figure 3 presents the combined transmission spectrum of a porous alumina membrane in a wide optical range. There are several major regions, and this allows us to make the complete analysis of the porous membrane properties.

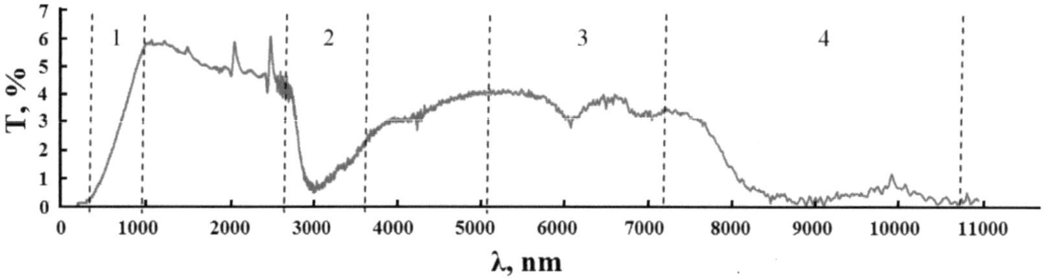

Figure 3. Transmission spectrum in a wide wavelength range. Different regions of spectrum characterize parameters of membranes: 1 – the average size and the dispersion of the pore diameter, 2 – the Al_2O_3 modifications, 3 – the thickness of membranes, 4 – anions of the used acid and the concentration of glycerol in the electrolyte.

Such spectra do not allow one to determine correctly the geometric parameters of a membrane. We can only state that the transmission spectrum shape corresponds to the geometric parameters of membrane determined by SEM.

Thus, the experimental data generalization gives a possibility to determine the average size and the dispersion of the pore diameter by measuring the UV–visible transmittance spectrum. The presence of different regions in the IR range of the transmission spectrum allows one to characterize the thickness of the membrane, the presence of residual products, as well as to identify the membrane material as alumina.

The RBS technique was used for in situ determination of the optimal geometry of the experiment. First, the position of a semiconductor detector at which the yield of the backscattered ions is uniform and minimal was found. This case corresponds to the situation when the pores in the target are through. It is worth noting that the above-mentioned ion-beam method of controlling the quality of nanoporous membranes structures is nondestructive.

The RBS spectra (figure 4) show the dependence of the number of passed particles on the slope angle of the target for the membrane containing pores of about 20 nm in diameter.

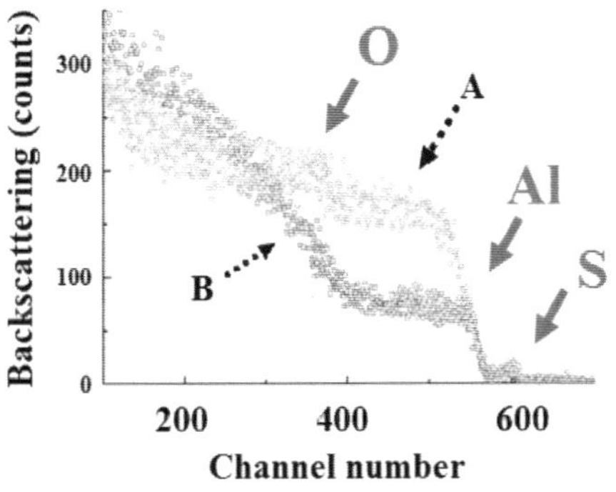

Figure 4. The energy spectrum of backscattered He+ ions with the energy of 1.7 MeV for the scattering angle of 120°. *A* – the structure of aluminum without etching, *B* after electrochemical etching [8]

The intensity of the passing beams is nearly constant within 2.5°. A halving in the signal corresponds to a 3° slope of the target. This proves that the charge accumulation on the inner face of the nanoporous membrane results in the guiding effect. The maximum transmission factor is measured to be 0.625.

4. Conclusions

Thus, the technology of producing self-forming porous membranes has been proposed. The porous membrane is a system of ordered nanoscale (pore diameter of 20 nm or more) alumina capillaries with an aspect ratio of 500. It was established that the optical spectroscopy of nanoporous membranes in a wide wavelength range (from 200 nm to 15 microns) provides a rapid characterization of their basic geometrical parameters including the membrane thickness, the average pore size and the pore size distribution. It is shown that the nanoporous membranes based on alumina containing a nanoscale capillaries system provide the transport of high-energy flow of helium ions with energy of 1.5–2 MeV.

Acknowledgments

The reported study was supported by RFBR, research project No. 14-08-31609 mol_a and by the Russian Ministry of Education, task (goszadanye) № 16.2112.2014/K (in the technological part, as well as in the part concerning the influence of preparation conditions on the structure of porous layers).

References

[1] Eftekhari A 2008 *Nanostructured Materials in Electrochemistry* (Weinheim: Wiley-VCH Verlag GmbH & Co. KGaA)

[2] Piao Y, Lim H, Chang J Y, Lee W-Y and Kim H 2005 Nanostructured materials prepared by use of ordered porous alumina *Electrochimca Acta* **50** 2997

[3] Gracheva I E, Moshnikov V A, Karpova S S and Maraeva E V 2011 Net-like structured materials for gas sensors *Journal of Physics: Conference Series* **291** 012017

[4] Foss C A, Hornyak G L and Stocketr J A 1993 Optically Transparent Nanometal Composite Membranes *Advanced Materials* **5** 135

[5] Afanas'ev A V, Il'in V A, Moshnikov V A, Sokolova E N and Spivak Yu M 2011 Synthesis of nanoporous and microporous structures by the electrochemical methods *Biotekhnosfera, in Russian* Issue 1–2 pp 20–26

[6] Muratova E N, SpivakYu M, Moshnikov V A, Petrov D V, Shemukhin A A and Shimanova V V 2013 Influence of technological parameters of nanoporous Al_2O_3 layers' preparation on their structural characteristics *Glass Physics and Chemistry* **39** 320

[7] Matyushkin L B and Muratova E N 2013 Investigation of the optical properties of nanoporous membranes based on Alumina *Smart Nanocomposites* **4** 86

[8] Shemukhin A A and Muratova E N 2014 Investigation of transmission of 1.7-MeV He^+ beams through membranes of porous Alumina *JETP Letters* **40** 219

Electrophysical properties of PCM-materials in crystalline and amorphous states

A V Gaydamaka, O A Martynova, E V Vladimirskaya and V E Gasumyants

Saint Petersburg Polytechnic University, 29 Polytechnicheskaya str., St. Petersburg 195251, Russia

E-mail: vgasum@yandex.ru

Abstract. The temperature dependences of the resistivity, thermopower, and Nernst coefficient were experimentally investigated on the samples of phase change materials with a composition of $Ge_{0.15}Sb_{0.85}$ in both the amorphous and crystalline states. It was observed that the crystalline state is characterized by a metallic type of the thermopower temperature dependences, while the resistivity increases slightly with decreasing temperature and the Nernst coefficient has a positive value and demonstrates the complicated temperature dependence. The conductivity in the amorphous state has an activated character with two different activation energies in low and high temperature ranges; the temperature of the transition between these ranges corresponds to the point of a change of thermopower sign. The results obtained point to a need to take into account a contribution of two types of carriers in the conduction process in both phase states.

1. Introduction

Phase change materials (PCMs) are the materials that exist in at least two structurally different solid phases, namely, in the amorphous and crystalline states. Usually, the amorphous and crystalline phases demonstrate very different optical and electrical properties originated from a large difference in the crystalline structure. This can be used in technological applications for storing information, if it is possible to change repeatedly the state of a material between the two phases and both phases are stable at operating temperatures. Nonvolatile memory electrical devices based on PCMs cause now an increasing interest, in particular as a possible alternative to the standard silicon flash memory [1,2].

A potential possibility to use PCMs for practical applications in fields of optical memory cells and phase-change recording was indicated a long time ago [3], but its realization was hampered due to the fact that the previously known PCMs were characterized by a rather slow crystallization speed. However, an interest to this class of materials increased greatly after discovering fast switching alloys based on the pseudobinary line between GeTe and Sb_2Te_3 [4]. As a result, at present the compounds of this family ($Ge_2Sb_2Te_5$, $GeSb_2Te_4$, GeTe, Sb_2Te_3 and others) are considered as the most promising materials for applications in the field of information storage. Moreover, PCMs have already been successfully used to create rewritable DVD and Blu-ray discs [1], as well as a prototype of a multi-layer memory cell based on the $Ge_2Sb_2Te_5$ compound [5].

In contrast to wide development of applications, the mechanism of a rapid phase change, especially crystallization from the amorphous phase, is still not fully understood in spite of a lot of investigations using different techniques (see, for example, [1, 6–8]). As a result, there is no consensus on the fundamental properties and parameters of the charge-carrier system. The questions on mechanisms of modification of those parameters under different kinds of influence, the possible ways of their purposeful variations are still under discussion. The main reason for this situation is a complicated or unusual energy structure of these materials differing essentially from that of "classical" materials. This circumstance prevents using the well-known models of the energy spectrum as a tool for an analysis of the experimental results on the electron transport. On the other hand, understanding of the transport properties and how they can be affected by different treatments of materials, the degree of disordering, electron-phonon and electron-electron interactions are the problems of fundamental importance for further progress in applications of the considered materials. Nevertheless, the experimental data on the transport coefficients behavior in PCMs are fragmentary; moreover, some of these coefficients have not been investigated yet.

For the above reasons, this work is devoted to the experimental study of the transport coefficients (the resistivity, thermopower, Nernst coefficient) on the $Ge_{0.15}Sb_{0.85}$ samples in both the amorphous and crystalline states.

1. Experiment

The $Ge_{0.15}Sb_{0.85}$ samples were obtained by the magnetron spattering, the samples size was approximately 10×10 mm^2. They were grown in an amorphous phase on the Si/SiO_2 substrate and then crystallized by a slow heating in one case or by a fast heating (at a rate of about 1° C/sec.) in the second case to test the reproducibility of the results. The sample thicknesses were 60 and 600 nm. The sample surface topology was studied by a Dimension 3100 (Veeco) AFM in the semi-contact mode at room temperature using the RTESP probes with the curvature radius of 10 nm.

The temperature dependences of resistivity, ρ, were measured by the van der Pauw method that allowed us to solve the problem of an arbitrary shape of the samples. The thermopower, *S*, was measured by the differential method relative to the copper electrodes and then calculated by correcting for the absolute thermopower of copper. The temperature difference between two edges of the sample was about 2 K. The Nernst coefficient, *Q*, was measured in the constant reversible magnetic field of 1.8 T in a home-made cryostat. In order to suppress spurious voltage, the Nernst signal was calculated as a half-sum of signals measured at different field directions. The temperature difference applied to the sample was about 10 K. All the transport coefficients were investigated in the temperature range of $T = 77–300$ K.

2. Results and discussion

To characterize the sample quality the AFM studies were carried out. Surface topologies for 60 nm $Ge_{0.15}Sb_{0.85}$ samples in the amorphous and crystalline states are shown in figure 1. All the samples have a granular structure with a particle size varying from 10 to 40 nm. The crystalline samples are characterized by a more uniform particle size than the samples of the amorphous phase. At the same time, the surface of all the samples is quite smooth; the surface irregularity is within limits of 10 nm. In the crystalline phase the height variation is a little less than in the amorphous samples for both the thicknesses investigated. Thus, all the sintered samples were of a high quality.

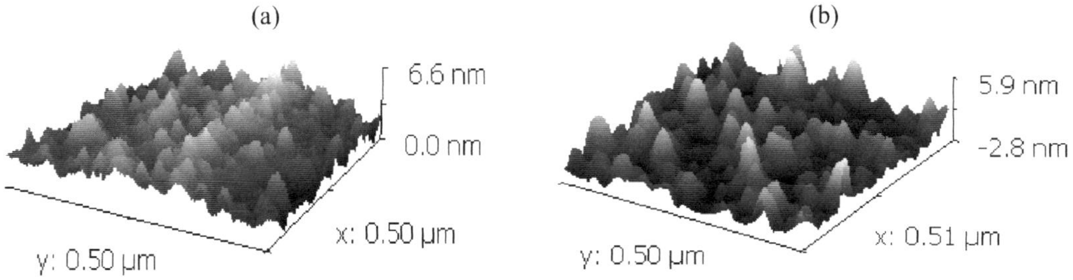

Figure 1. Surface topology of $Ge_{0.15}Sb_{0.85}$ samples (thickness of 60 nm) in crystalline (a) and amorphous (b) states.

The resistivity temperature dependences for the samples of 600 nm thickness are shown in figure 2. These dependences for each sample were measured twice in the van der Pauw configuration, and the resistivity values were calculated taking into account the correction factor determined by the deference between the values obtained by repeating measurements after changing the configuration of contacts [9].

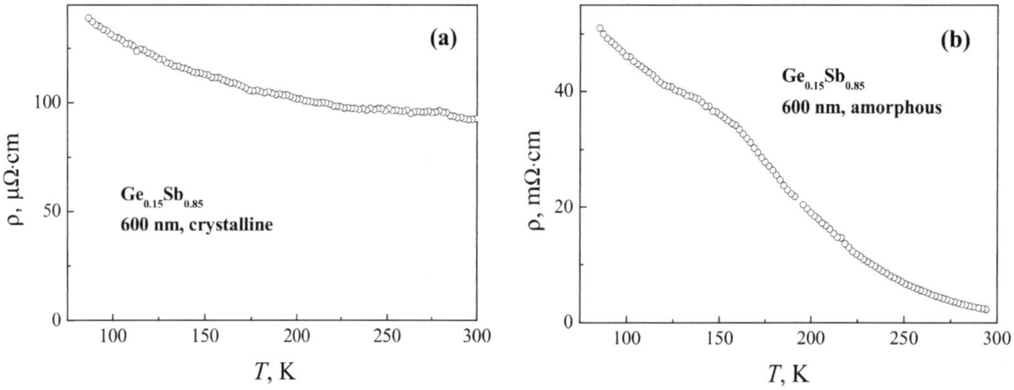

Figure 2. Resistivity vs temperature for the 600 nm thickness $Ge_{0.15}Sb_{0.85}$ samples in crystalline (a) and amorphous (b) states.

The absolute values of resistivity at $T = 300$ K for the samples of the amorphous state were found to be 25–30 times higher than for the crystalline samples. As for the temperature dependence, in the crystalline state the resistivity increases with decreasing temperature but rather slightly (for the sample of 600 nm thickness the resistivity increases by 36 % when temperature decreases from 300 K to 100 K, see figure 2(a)). In the amorphous phase we have observed a strong rise of the resistivity with decreasing temperature (see figure 2 (b)) that indicates an activation type of the conduction process to be realized in this sample state.

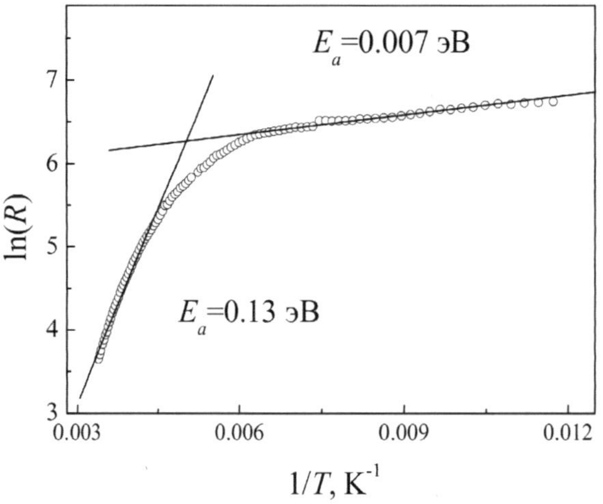

Figure 3. Temperature dependence of the resistivity for the 600 nm thickness amorphous $Ge_{0.15}Sb_{0.85}$ sample in Arrhenius coordinates.

To determine the quantitative characteristics we have plotted the $\rho(T)$ dependence for the amorphous samples in Arrhenius coordinates (see figure 3). The presented data clearly indicate that the conduction process in different temperature ranges is characterized by the two different activation energies, E_a. At high temperatures ($T > 200$ K) the $\rho(T)$ dependence corresponds to $E_a = 0.13$ eV, while at low temperatures ($T < 150$ K) $E_a = 0.007$ eV.

Figure 4 shows the temperature dependences of the thermopower for the amorphous and crystalline samples of 60 nm thickness. In the crystalline state the absolute values of the thermopower are extremely small (about 4–5 μV/K at $T = 300$ K); the $S(T)$ dependence is of a metallic type, i.e. thermopower decreases linearly with decreasing temperature. In the amorphous state the S values are also quite small (they do not exceed 120 μV/K), while the temperature dependence of the thermopower has a complicated character demonstrating a transition from positive values at $T = 130$–300 K to negative ones at $T < 125$ K. Note, that the temperature of a transition to negative thermopower values is close to that of slope change for the $\rho(T)$ (see figure 2(b)).

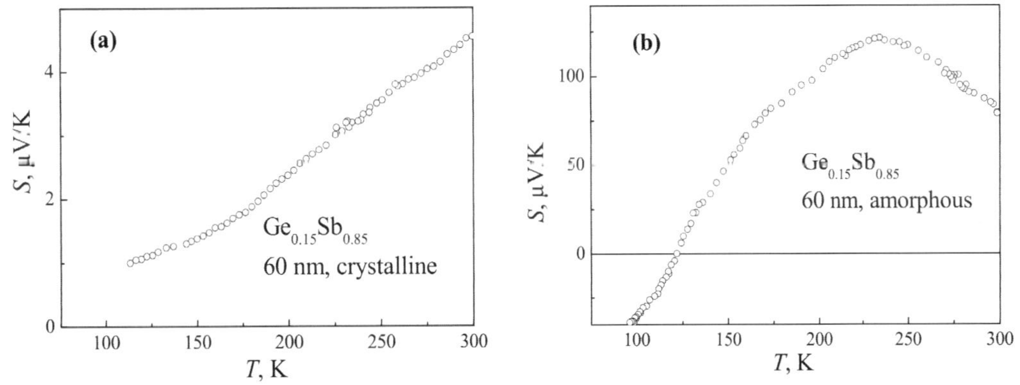

Figure 4. Thermopower vs temperature for the 60 nm thickness $Ge_{0.15}Sb_{0.85}$ samples in crystalline (a) and amorphous (b) states.

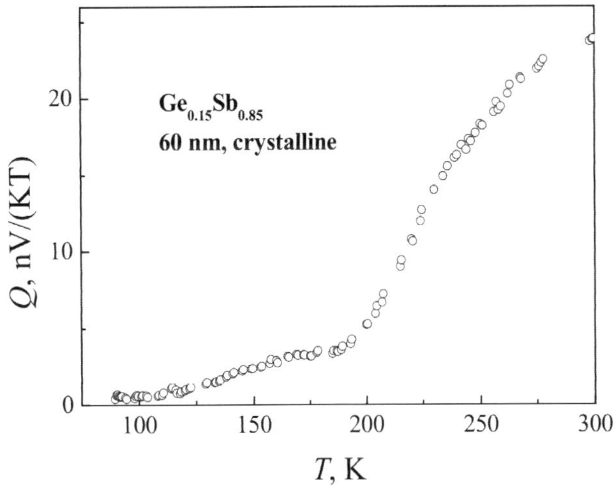

Figure 5. Nernst coefficient vs temperature for the 60 nm thickness crystalline $Ge_{0.15}Sb_{0.85}$ sample.

For the crystalline sample of 60 nm thickness we have also measured the temperature dependence of the Nernst coefficient. Note, that to the extent of our knowledge no data on this coefficient behavior in PCMs of the studied composition have been published to date. The results obtained are shown in figure 5. The Nernst coefficient at $T = 300$ K is positive and equal to about 25 nV/(KT). The $Q(T)$

dependence has two different regions. As temperature decreases down to 200 K the Q value falls sharply, at lower temperatures it decreases much slighter. At $T < 100$ K the Q values become extremely low (about 0.5 nV/(KT)).

Let us discuss briefly the results obtained. As for the amorphous state, it is obvious that to describe the conduction process one should take into account a contribution of two types of carriers. According to the resistivity and thermopower data at low temperatures the main charge carriers are electrons characterized by an activation energy of $E_a = 0.007$ eV, while at high temperatures (above 150–200 K) they are holes with an activation energy of $E_a = 0.13$ eV. The data on the crystalline state are rather contradictive. The type of the thermopower temperature dependence and very small absolute values of it give evidence of a strong degeneracy of the electron gas. On the other hand, the $\rho(T)$ dependence is not a metallic-like. As for the Nernst data obtained, in case of one type of charge carriers they can be explained only by a change in a dominant scattering mechanism with decreasing temperature. However, the resistivity and thermopower data testify against such an assumption. Thus, we believe the results obtained for three transport coefficients to be evidence that the existence of two types of charge carriers is characteristic of both the amorphous and crystalline states of PCMs.

3. Conclusion

Thus, by the method of magnetron sputtering we have sintered the high quality thin film samples of the $Ge_{0.15}Sb_{0.85}$ composition with thicknesses of 60 and 600 nm. We have studied experimentally the temperature dependences of the resistivity and thermopower for both the amorphous and crystalline states, as well as the Nernst coefficient for the crystalline state. The crystalline state is characterized by a metallic-like thermopower temperature dependence, while the resistivity increases with decreasing temperature and the Nernst coefficient having small positive absolute values demonstrates a rather complicated temperature dependence. In the amorphous state the conductivity has an activated character with different activation energies at low and high temperatures; the thermopower has positive values at high temperatures, but negative ones below $T = 125$ K. The qualitative analysis of the results obtained indicates that describing the conduction process in both the amorphous and crystalline states requires taking into account a contribution of charge carriers of different types.

Acknowledgments

The study was supported by the grant of the St. Petersburg government in the field of scientific and technical activities. Authors thank Dr. P. Brunkov for performing the AFM measurements. This part of work was made on the equipment of the Joint Research Centre «Material science and characterization in advanced technology» (Ioffe Institute, St. Petersburg, Russia).

References

[1] Wuttig M and Yamada N 2007 *Nature Materials* **6** 824
[2] Raoux S, Wełnic W and Ielmini D 2010 *Chemical Reviews* **110** 240
[3] Ovshinsky S R 1968 *Physical Review Letters* **22** 1450
[4] Yamada N, Ohno E, Akahira N, Nishiuchi K, Nagata K and Takao M 1987 *Japanese Journal of Applied Physics* **26-4** 61
[5] Chen Y C 2009 *Phase Change Materials: Science and Applications* ed S Raoux and M Wuttig (New York: Springer) pp 331–354
[6] Kimura S, Tanaka Y, Kohara S and Takata M 2012 *Materials Science and Technology* ed S D Hutagalung (Rijeka: InTech) pp 259–275
[7] Lencer D, Salinga M and Wuttig M 2011 *Advanced Materials* **23** 2030
[8] Redaelli A, Pirovano A, Benvenuti A and Lacaita A L 2008 *Journal of Applied Physics* **103** 111101
[9] van der Pauw L J 1958 *Philips Research Reports* **13** 1

Journal of Physics: Conference Series **586** (2015) 012010 doi:10.1088/1742-6596/586/1/012010

Research of vacancy defect formation on the surface of two-dimensional boron sheets

E V Boroznina, T A Zhiganova and S V Boroznin

Volgograd State University, 100 University Ave., Volgograd 400062, Russia

E-mail: extrajenya@bk.ru

Abstract. Mechanism of vacancy defect formation on the surface of two types of BS: triangular BS (TBS), α-sheet (αBS) have been calculated within the model of molecular cluster with the use of quantum chemical MNDO scheme. The process of atomic vacancy formation of BS has been modeled by step-by-step abstraction of one central boron atom. Incremental method allowed us to build energy curves for vacancy formation process. The process of vacancy migration on the BS surface has also have been investigated, and more probable paths of migration have been found.

1. Introduction

In recent decades boron has become an element of vital importance since both boron and its compounds are widely used in many fields of modern technology. Nowadays, many research teams and laboratories are engaged in the researches connected with boron and its compounds. It has been found that boron atoms are capable of forming both ionic and covalent bonds. They can form chains, frames, lattices etc. Recently, scientists are interested in two-dimensional boron sheets (BS) [1]. The synthesis of novel two-dimensional (2D) materials has attracted considerable interest because of their numerous unique properties. While the production of BS remains a challenge it is important to theoretically investigate the possible fabrication methods and physical-chemical properties of it. The elemental boron holds a unique place among the elements of the periodic table by having the most varied polymorphism. So, the structure of BS can be very diverse. Concerning the extensive theoretical studies of boron clusters it can be established that the most stable BS are the triangular BS (TBS) and α-boron sheet (αBS) [2–4]. The examples of TBS and αBS are shown in figure 1.

2. Vacancy defect formation

It is known that the point surface defects affect the physical-chemical properties of materials. In this paper we present the investigation of the mechanism of vacancy defect formation on the surface of two types of BS (TBS and αBS) and the transport properties of vacancies. The calculations are carried out within the model of molecular cluster with the use of quantum chemical MNDO scheme [5]. The length of B-B bonds is equal to 1,67 Å. These data were calculated in earlier theoretically studies [2–4]. The unsaturated cluster boundary chemical bonds have been completed by hydrogen atoms.

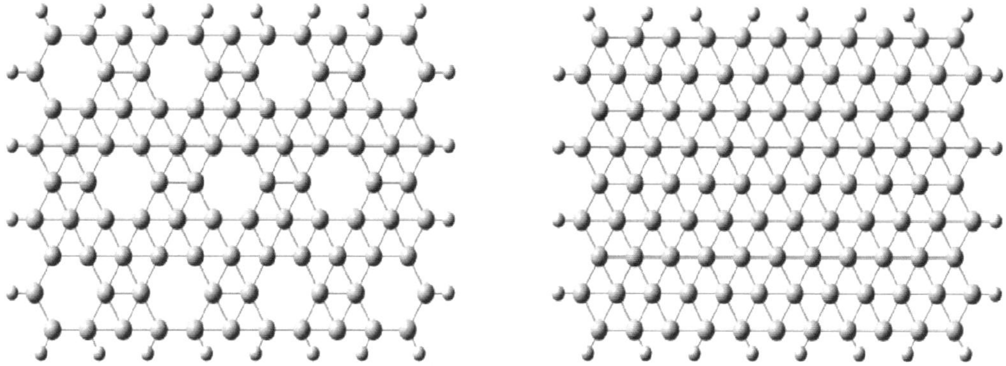

(a) (b)

Figure 1. (a) α-sheet of boron atoms (αBS); (b) triangular boron sheet (TBS).

To model the process of vacancy formation a central B-atom was moved from the surface of BS along the perpendicular to the BS with a step of 0.1 Å until the moment of its separation.

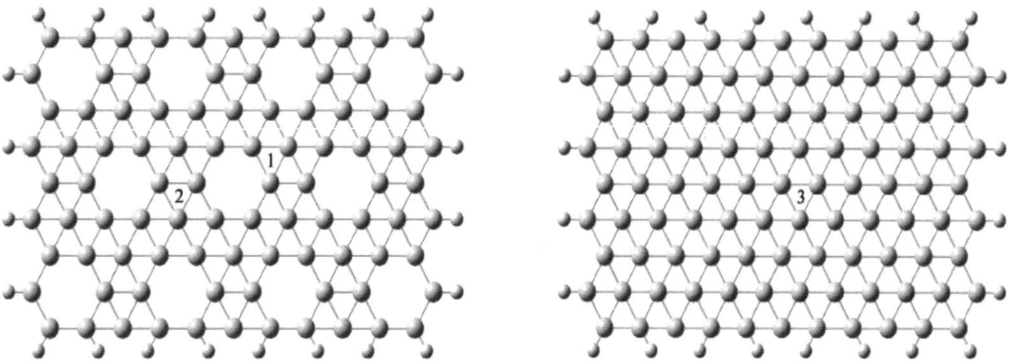

(a) (b)

Figure 2. The clusters of boron sheets: (a) α-boron sheet, (b) triangular boron sheet. 1, 2, 3 denote the locations of vacancy defect

As a result, we obtained a potential energy profile of the vacancy formation process shown in figure 3. The analysis of the energy curve allows us to conclude that the boron atom in all cases is gone in the energy minimum (figure 3). The minimum energy E_v is situated 0.8 Å apart from the atom B to planar surface of the sheet. In this case the length of B-B bond changes from 1.67 Å to 1.85 Å. It is obvious that such position of surface the atom B is stable too. In case of the curve 1 the energy hump of about 22 eV has been obtained. It can be explained by the analysis of the atom distribution in each option of atom location. In the position 1 there are less neighbors than in the positions 2 and 3. So, the gravity forces between atoms are stronger and the repulsive forces are weaker. The high value of this energy hump can be explained by using the cluster model. It is well known that the cluster models often give the correct description of the process but the values of some parameters may be higher than the experimental data.

Journal of Physics: Conference Series **586** (2015) 012010 doi:10.1088/1742-6596/586/1/012010

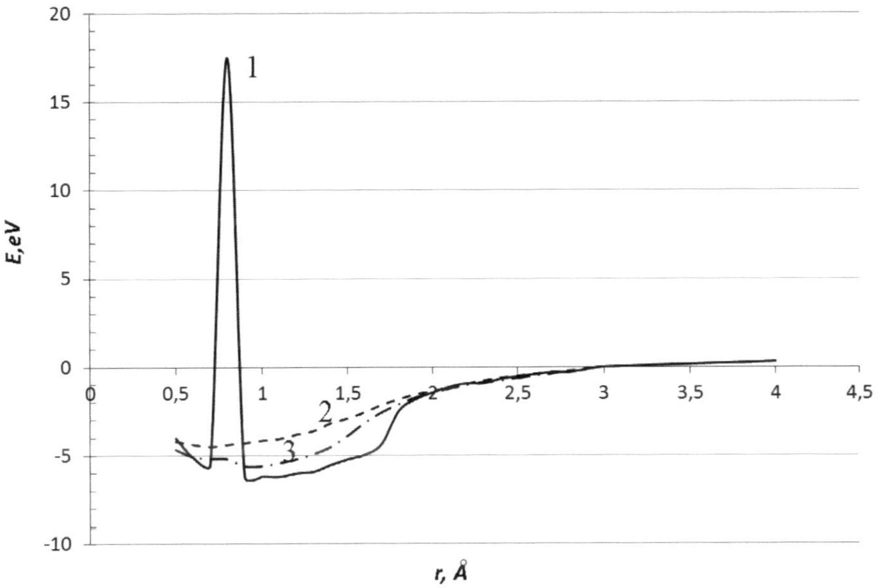

Figure 3. Energy curves for vacancy formation process for boron sheets. Curves 1, 2 correspond to positions 1, 2 for αBS (see figure 2 (a)). Curve 3 corresponds to TBS (position 3 in figure 2 (b)). r is the distance from BS surface, E is the energy of the system.

The analysis of the electronic structure and energy characteristics showed that a defect changes the width of the energy gap. The width of the energy gap increases in BS with a defect.

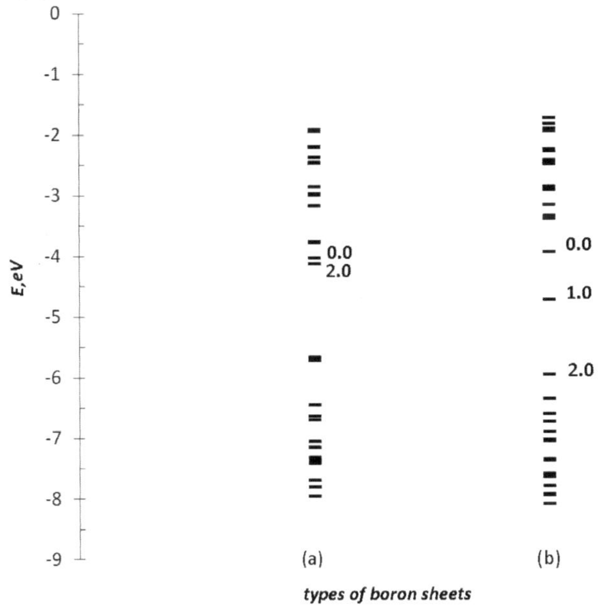

Figure 4. Single electron energy spectra of:
(a) pure α-boron sheet; (b) α-boron sheet with vacancy defect.

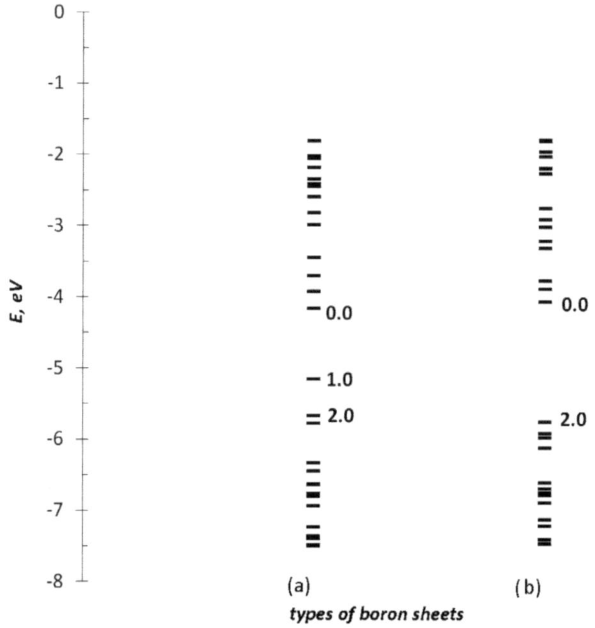

Figure 5. Single electron energy spectra of:
(a) pure triangular boron sheet; (b) triangular boron sheet with a vacancy defect

3. Vacancy migration

Next step of our research was to study the energy characteristics of vacancy defect migration on the surface of BS. The process was modeled by approaching the neighboring boron atom to the place of defect location along the virtual B-B bond. Thus, the boron atom of BS has two degrees of freedom and can move within the surface of the layer and deviate from it. Geometric parameters of other two atoms near the vacancy were optimized. Therefore, it seemed that the vacancy moves in the direction opposite to the movement of the migrating atom. We considered one type of migration vacancy for TBS and four types of migration vacancy for the αBS (figure 6).

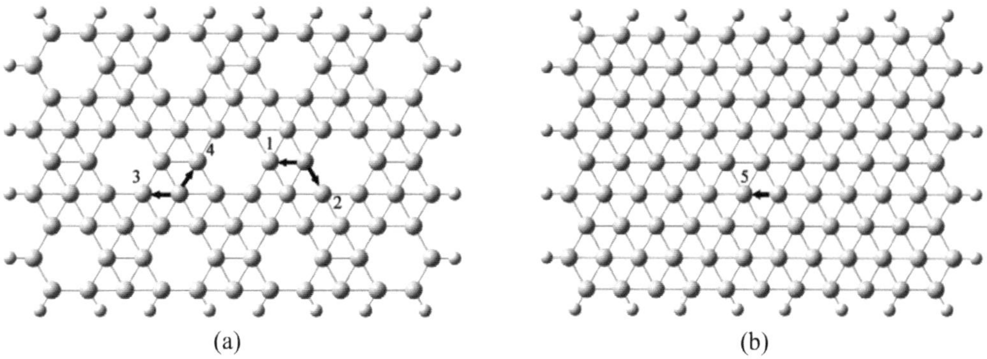

Figure 6. Vacancy migration: (a) α-boron sheet, (b) triangular boron sheet.

As a result of successive approximations of the boron atom we have got the profile of potential energy of vacancy migration. The analysis of energy curves (figure 7) shows that the vacancy migration is possible along the path 2 in αBS and along the path 5 in TBS. The energy curves for these paths almost overlap and potential energy changes a little on the way of the defect. In other cases (paths 1, 3, 4) there is a strong growth of potential energy of vacancy migration, so we can conclude that this process is energetically unfavorable and improbable.

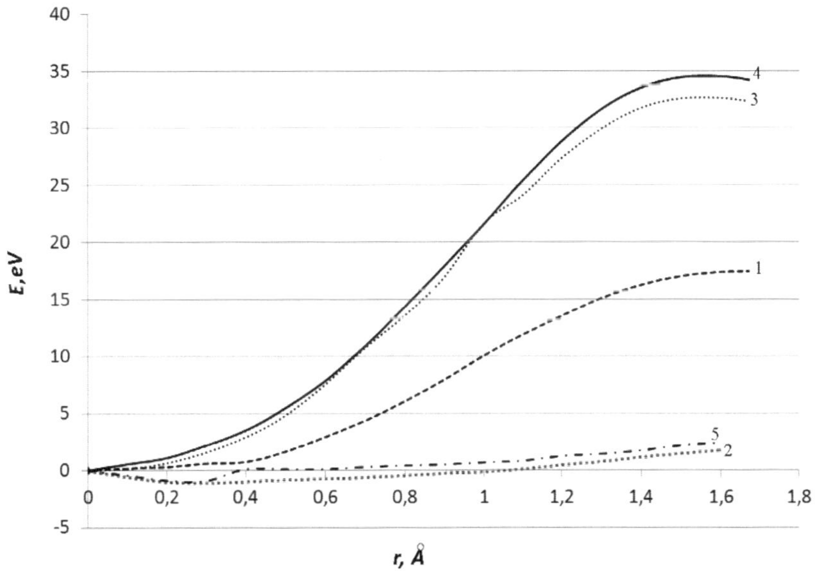

Figure 7. Energy curves for vacancy migration in an α-boron sheet (1-4) and a triangular boron sheet (5). The curves 1, 2, 3 and 4 correspond to migrations to the positions 1, 2, 3 and 4 respectively. *r* is the distance from the boron atom location to the next step, as it is shown in figure 6, Å. $r = 0$ corresponds to the location of the boron atom before the vacancy migration starts. *E* is the energy of the system.

4. Conclusion

Research of vacancy defect formation and the transport properties of vacancy on the surface of two-dimensional boron sheets have been investigated. The BS with a single vacancy defect is stable. The width of the energy gap increases in the defected BS.

The process of vacancy migration on the surface of TBS is possible in any direction unlike the αBS where it can be implemented along one path only.

References

[1] Xiaobao Y, Yi D and Jun N 2008 *J. Phys. Rev.* **77** 041402
[2] Lau K C 2007 *J. Phys. Chem.* **111** 2906
[3] Evans M H 2005 *Phys. Rev.* **72** 045434
[4] Kunstmann J 2006 *Phys. Rev.* **74** 035413
[5] Dewar M J S 1977 *Theoret. Chem. Acta* **46** 089104

Study of temperature and pressure dependences of electric properties of amorphous Ag-Ge-As-S chalcogenides containing carbon nanotubes

K V Kurochka, N V Melnikova, V E Zaikova and Ya Yu Volkova

Ural Federal University, 19 Mira str., Ekaterinburg, 620002, Russia

e-mail: kirill.k.v@yandex.ru

Abstract. Frequency and temperature dependences of impedance of $AgGe_{1+x}As_{1-x}(S+CNT)_3$, x = 0.4; 0.5; 0.6 were measured. There are maxima on frequency dependences of tangent of loss, which can be connected with relaxation processes inside the sample (the maximum at high frequencies) and the second maximum can be connected with processes occurring on the sample-electrode border. Activation energy of all synthesized materials was estimated to be near 0.5 eV. The relaxation of electric resistance at fixed pressures within the pressure range of 30 to 50 GPa was observed. It is established that the resistance decreases exponentially with time within this range of pressures.

1. Introduction

Multicomponent silver chalcogenides with a silver ion conductivity have been known as promising materials for scientific and applied purposes. They can be applied in a wide range of devices such as memory cells [1], solid state batteries, different kind of sensors etc. Materials of the Ag-Ge-As-S system have a significant share of ionic component in conductivity [2, 3]. In order to obtain materials with a higher ionic conductivity component, glassy silver chalcogenides $AgGe_{1+x}As_{1-x}(S+CNT)_3$ with the addition of carbon nanotubes (CNT) were synthesized. The obtained materials have the ionic conductivity share not less than 99 percent [4].

This paper is devoted to the study of temperature dependences of electric properties at ambient pressure and resistance relaxation at fixed pressure in the pressure range of 30 to 50 GPa in glassy $AgGe_{1+x}As_{1-x}(S+CNT)_3$ (x = 0.4; 0.5; 0.6).

2. Materials and methods of study

The synthesis of $AgGe_{1+x}As_{1-x}(S+CNT)_3$ (x = 0.4; 0.5; 0.6) with about 7 additional atomic percent of CNT was carried out by means of alloying initial components in the quartz containers evacuated to residual pressure of 10^{-4} Pa, filled with a superpure argon to 0.5 10^5 Pa. The initial components were the chemical reagents of high purity and nanotubes added to the blend as a composite with dehydrated sulfur.

The materials were investigated by an X-ray diffractometer Shimadzu XRD 7000 (monochromatic radiation CuKα). An Auriga CrossBeam Zeiss scanning electron microscope providing the X-ray microanalysis (EDS) was used to confirm the homogeneity of glasses as well as to estimate the atomic percent ratio of elements. The synthesis and properties of materials at room temperature are described in detail in [4, 5].

Electrical properties in the temperature range of 300 to 383 K were investigated by impedance spectroscopy in the frequency range of 1 Hz to 32 MHz in the cells with graphite electrodes. Measurement and analysis of impedance was performed by a universal frequency response analyzer Solartron 1260A.

Experiments under high pressure (within the range of 15 to 50 GPa) were performed using powder samples in rounded cone – plane type high pressure cell (HPC). Conductive synthetic carbonado diamonds were used as a material for anvils. These diamonds are good conductors, which allow measuring the electrical properties of materials placed in the pressure cell [6]. The measurements of the electrical resistance of materials in HPC were performed using an Agilent 34970A.

3. Results and Discussion

Since the materials under study have the ion component of conductivity of about 99 percent, applying an external electric field results in observing the polarization phenomena near electrodes. In case of alternating current these phenomena appear much weaker than at direct current and depend on the frequency of alternating current and the electrodes' material.

Thus, the studies of temperature dependences of impedance in the frequency range of 1 Hz to 32 MHz (figure 1) in the cells with the electrodes blocking the ion component of conductivity (graphite) were performed. As it can be seen in figure 1, these materials are characterized by two well separated areas on the frequency dependence (in the whole temperature range from 300 to 383 K): the high-frequency one characterizing properties of the sample, and the low-frequency one characterizing the influence of the electrode impedance.

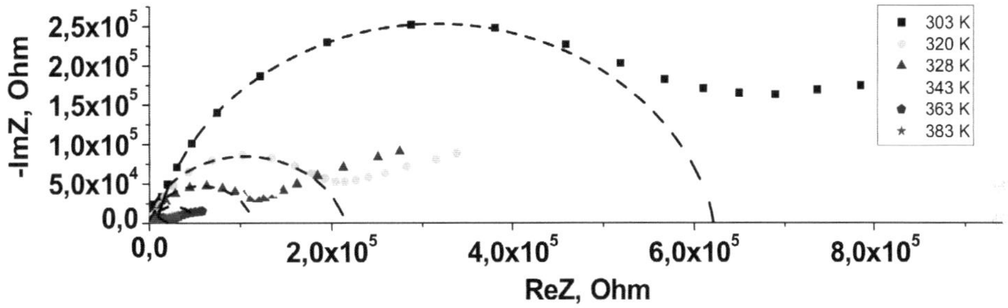

Figure 1. Hodographs of impedance of $AgGe_{1.4}As_{0.6}(S+CNT)_3$ in the cell with graphite electrodes in the temperature range of 300 to 383 K. Dash lines indicate the circle approximation of high frequency parts of hodographs which characterize properties of the sample.

There are relaxation maxima on the frequency dependences of tangent of dielectric loss for the sinthesized materials (figure 2 (b)). The maximum at high frequencies is associated with the relaxation processes in the sample, and the estimated relaxation time is typical for the ion-relaxation polarization. The appearance of the maximum at low frequencies can be connected with the processes occurring on the sample-electrode border [7]. It is also confirmed by the dramatic increase of permittivity determined by the double dielectric layer capacitance at low frequencies (figure 3). The maxima shift towards higher frequencies with increasing temperature as it is shown in figure 2 (a) for $AgGe_{1.5}As_{0.5}(S + CNT)_3$. The same dependence of dielectric loss tangent was also observed in the materials without CNT. The high frequency peak in the frequency dependence of tangent of dielectric loss shifts toward higher frequencies with increasing temperature because the time required for the completion of this relaxation process decreases with increasing temperature that results from weakening of bonds between ions, increasing number of mobile ions and easier electron transport. The frequency at which polarization occurs on the electrodes because of accumulation of space charge on the sample-electrode border also becomes higher with increasing temperature. Adding nanotubes also leads to the fact that the silver ions can move inside CNTs [8] that increases their mobility resulting in the decrease of the relaxation time.

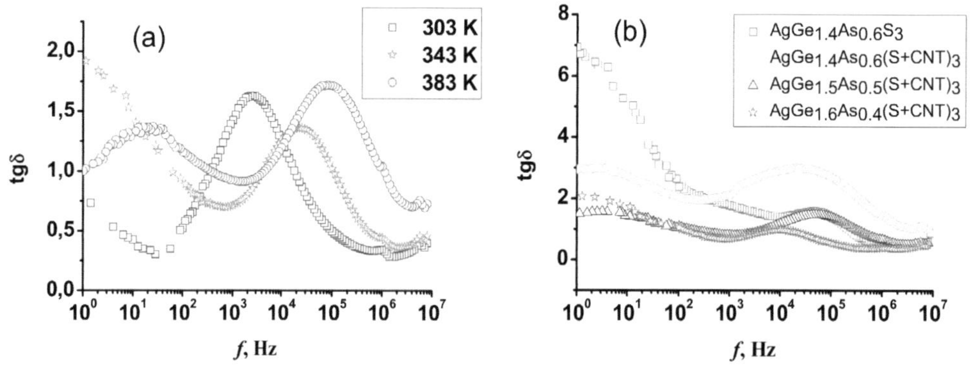

Figure 2. Frequency dependences of tangent of dielectric loss (tgδ): (a) for $AgGe_{1.5}As_{0.5}(S+CNT)_3$ ($x = 0.5$) at different temperatures; (b) – for $AgGe_{1.4}As_{0.6}S_3$ and $AgGe_{1+x}As_{1-x}(S+CNT)_3$ ($x = 0.4$; 0.5; 0.6) at 363 K.

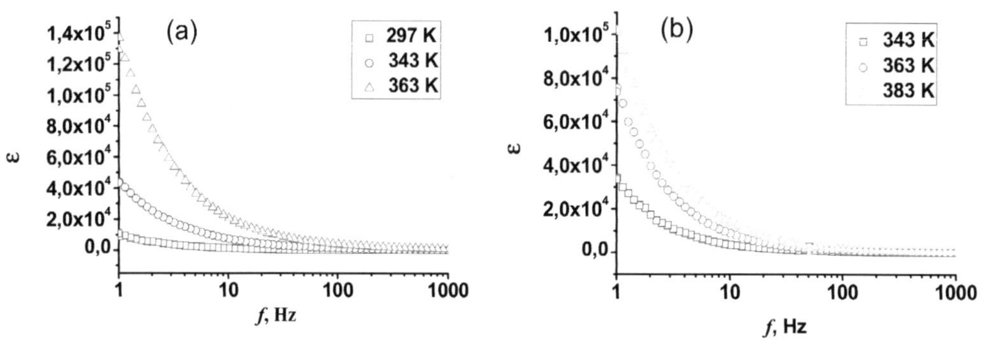

Figure 3. Frequency dependences of permittivity at different temperatures: (a) for $AgGe_{1.4}As_{0.6}(S+CNT)_3$ ($x = 0.4$); (b) for $AgGe_{1.4}As_{0.6}S_3$.

The activation energy of all studied materials was estimated based on the temperature dependences of resistivity. In the temperature range of 300 to 383 K it is equal to 0.51 eV for the compounds containing CNT with $x = 0.4$; 0.5 and to 0.54 eV for $x = 0.6$ (table 1).

The values of activation energy we obtained are in a good agreement with the values obtained in [2] for $AgGe_{1+x}As_{1-x}S_3$ ($x = 0.4$; 0.6) compounds.

In the $AgGe_{1+x}As_{1-x}S_3$ and $AgGe_{1+x}As_{1-x}(S+CNT)_3$ compounds the increase of resistivity and decrease of conductivity with time at ambient pressure were observed in other works [2, 4] that is typical for the materials with an ionic conductivity component. However, it can be assumed that with increasing pressure ion transport practically disappears at some pressure, and only electrons are the charge carriers. In fact, the investigation of resistivity relaxation at fixed pressures above 30 GPa confirms that the processes occurring in $AgGe_{1.4}As_{0.6}S_3$ and $AgGe_{1.4}As_{0.6}(S + CNT)_3$ are typical for the materials with only electron transport. The study of the electrical relaxation at fixed pressure in the range of 30 to 50 GPa showed that the resistance decreases exponentially with time for both materials (figure 4).

Table 1. The activation energy of compounds with/without CNT (including the data from [2])

Compound	Activation energy	Temperature range
$AgGe_{1.4}As_{0.6}(S+CNT)_3$	0.51	300 – 383 K
$AgGe_{1.5}As_{0.5}(S+CNT)_3$	0.51	300 – 383 K
$AgGe_{1.6}As_{0.4}(S+CNT)_3$	0.54	300 – 383 K
$AgGe_{1.4}As_{0.6}S_3$	0.40	300 – 383 K
$AgGe_{1.6}As_{0.4}S_3$	0.47 [2]	240 – 300 K

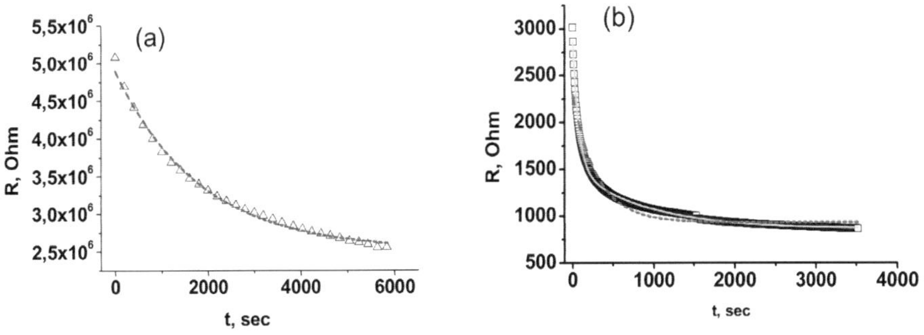

Figure 4. Time dependence of resistance at 46 GPa: (a) for $AgGe_{1.4}As_{0.6}S_3$ (b) for $AgGe_{1.4}As_{0.6}(S+CNT)_3$. The dash and solid lines show the exponential and two exponential approximations of experimental data respectively.

Exponential time dependence of the resistance can be approximated by the expression [9]:

$$R(t) = (R_0 - R_\infty)\exp(-\frac{t}{\tau}) + R_\infty, \tag{1}$$

where τ is the relaxation time, R_0 and R_∞ - the resistance at $t = 0$ and in the saturated state respectively.

However, the approximation of experimental data by (1) can be not satisfactory enough for the materials containing CNT. In this case, the approximation by an exponential function of the form:

$$R(t) = (R_0 - R_\infty)\exp(-\frac{t}{\tau_1}) + R_{\text{int}}\exp(-\frac{t}{\tau_2}) + R_\infty, \tag{2}$$

can be more exact (figure 4 (b)).

Then the total relaxation process is characterized by two different relaxation times, one of them (τ_1) describes the relaxation processes occurring at deformation of CNTs and resulting in increasing electron conductivity, while the second one (τ_2) may be connected with the increase in electrical conductivity during deformation and compaction of glass network with increasing static pressure. In this case the time τ, determined by the expression (1) is an average relaxation time.

The relaxation times for different values of fixed pressure can be estimated using the exponential approximation (1). The pressure dependences of the relaxation times for $AgGe_{1.4}As_{0.6}S_3$ and $AgGe_{1.4}As_{0.6}(S + CNT)_3$ compounds are shown in figure 5.

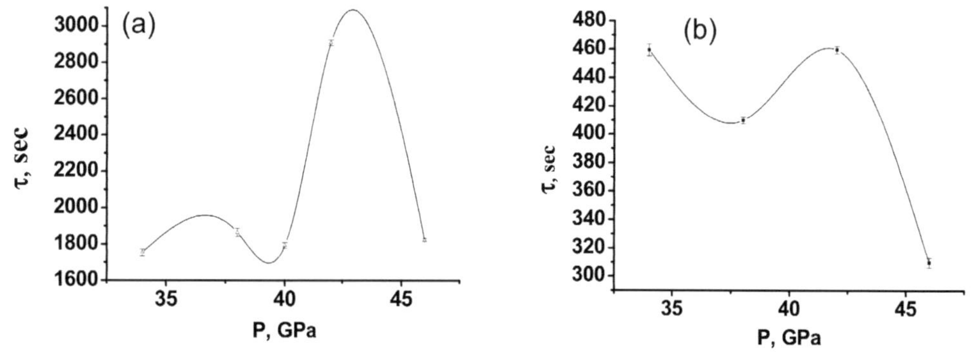

Figure 5. Variation of the average relaxation time defined by (1) with pressure: (a) for $AgGe_{1.4}As_{0.6}S_3$; (b) for $AgGe_{1.4}As_{0.6}(S+CNT)_3$.

As it can be seen in figure 5, introduction of CNTs into the glass network leads to increasing of electronic conductivity of glass [8], which results in a decrease of resistance and relaxation time under high pressure. Moreover, there are the specific features in the distribution of relaxation times near the pressure value of 42 GPa for both materials.

4. Conclusion

The temperature dependences of the electrical properties of $AgGe_{1+x}As_{1-x}(S+CNT)_3$ with $x = 0.4$; 0.5; 0.6 at alternating electric field within the frequency range of 1 Hz to 32 MHz and in the temperature range of 300 - 383 K have been studied. There are two peaks in frequency dependences of tangent of loss for all the materials. The first maximum observed at high frequencies is associated with the relaxation phenomena occurring in the sample. The second maximum observed at low frequencies may be associated with the phenomena at the sample-electrode border. Maxima shift towards higher frequencies with increasing temperature. The activation energy for all the materials studied was estimated based on the temperature dependences of resistivity. In the temperature range of 300 to 383 K it is equal to 0.51 eV for the compounds containing CNT with $x = 0.4$; 0.5 and to 0.54 eV for $x =$ 0.6.

The relaxation of the resistivity for $AgGe_{1.4}As_{0.6}(S+CNT)_3$ and $AgGe_{1.4}As_{0.6}S_3$ at fixed pressures in the range of 30 to 50 GPa has also been investigated. The resistance decreases exponentially with time for both materials. Analysis of pressure dependences showed that specific features in the distribution of relaxation times near the pressure value of 42 GPa are observed for both materials that can be associated with structural changes in glasses and the features of electron transport process.

Acknowledgments
The study was supported by the Russian Foundation for Basic Research, projects 13-02-00633, 13-02-96039-r_ural_a and by the Ural Federal University development program with the financial support of young scientists.

References

[1] Andrikopoulos K S, Arvanitidis J, Dracopoulos V, Christofilos D, Wagner T and Yannopoulos S N 2011 *Appl. Phys. Lett.* **99** 171911

[2] Kheifets O L, Shakirov E F, Melnikova N V, Filippov A L and Nugaeva L L 2012 *Semiconductors.* **46** 943

[3] Baranova E R, Kobelev V L, Kobeleva O L, Melnikova N V, Zlokazov V B, Kobelev L.Ya and Perfiliev M V 1999 *Solid State Ionics.* **124** 255

[4] Kurochka K V, Melnikova N V, Ustinova I S, Volkova Ya Yu and Alikin D O 2014 *J. Phys.: Conf. Ser.* **500** 192007

[5] Melnikova N V, Kurochka K V, Kheifets O L, Alikin D O, Kadyrova N I and Volkova Ya Yu 2014 *Book of Abstr. of XVII Int. Symposium Ordering in minerals and alloys* issue 17 **1** 203

[6] Babushkin A N 1992 *HPR* **6** 349

[7] Bordovsky G A and Kastro R A 2002 *Proceedings of Herzen RSPU* issue 4 **2** 7

[8] Stehlik S. et al. 2010 *J. Solid State Chem.* **183** 144

[9] Titus S S K, Skauragi H, Hayashi K, Kondo A and Shimakawa K 1996 *Journal of Non-Crystalline Solids* **198-200** 556

Intersubband absorption modulation in the GaAs/AlGaAs double tunnel-coupled quantum wells

D A Rybalko[1], M Ya Vinnichenko[1], L E Vorobjev[1], D A Firsov[1], R M Balagula[1], M M Kulagina[2] and A P Vasil'iev[2]

[1] Department of Physics of Semiconductors & Nanoelectronics, St. Petersburg Polytechnic University, 29 Polytechnicheskaya str., St. Petersburg 195251, Russia
[2] Ioffe Physical-Technical Institute, 26 Polytechnicheskaya str., St. Petersburg 194021, Russia

e-mail: dimarybalko91@gmail.com

Abstract. Structures with multiple doped double tunnel-coupled GaAs/AlGaAs quantum wells were grown and formed in mesa configuration. Positions of the energy levels have been verified with interband photoluminescence measurements. Temperature modulation of the intersubband absorption spectra was registered. Intersubband absorption modulation under transverse electric field was detected and explained.

1. Introduction

Recently, the fast frequency modulation of the radiation of quantum cascade lasers (QQL) was suggested for the use in communication [1]. Such modulation type increases the signal / noise ratio at heterodyne reception as compared with amplitude modulation [2] and allows the weatherproof optical communication over long distances. Frequency modulation of laser radiation can be realized if we embed system of tunnel-coupled quantum wells in QQL waveguide. Transverse (vertical) electric field applied in the growth direction changes the absorption coefficient of light polarized along the growth axis of the structure [3], and simultaneously changes the dielectric constant of the layer [4-7]. The change of layer refraction index leads to the change of radiation frequency. The change of intersubband light absorption in vertical electric field can be associated with two factors. The first one is the Stark shift of energy levels. In stepped and tunnel-coupled QWs the Stark shift can be rather strong (see theoretical works [8, 9]). Besides, transverse electric field can redistribute electrons between energy levels resulting in additional absorption modulation.

This work is devoted to the investigation of intersubband absorption modulation in double tunnel-coupled QWs GaAs / AlGaAs induced by the application of transverse electric field.

2. Sample and experiment

The structure was MBE grown on semi-insulating GaAs substrate and contained 100 pairs of double tunnel-coupled GaAs/AlGaAs QWs. The width of the quantum wells and alloy composition of the barriers were chosen in order to have three levels of dimensional quantization in the QWs system. The narrow and wide quantum wells have widths of 4.5 and 6.6 nm, respectively; the barrier $Al_{0,25}Ga_{0,75}As$ is 3 nm thick. The energy gap between the electronic levels e1 and e2 is about 30 meV. The energy of the e1-e3 transition is about 125 meV, energy of e2-e3 transitions is about 95 meV. Doping with silicon was performed in the central part of the wide quantum well (~2.5 nm) resulting in surface electron concentration of about $4{\cdot}10^{11}$ cm^{-2}. The highly doped (10^{18} cm^{-3}) GaAs layers were created above and under the QW layers in order to ensure effective electric field application. The thicknesses of these layers are 1.5 and 0.5 μm, respectively. Mesa was etched on the surface of structure at the stage of post-growth processing to ensure access to the lower contact layer. AlGaAs stop-layer above the lower doped layer was used to stop the selective etching of the structure.

It is well known that the normal incidence geometry of the experiment is not suitable for the intersubband absorption investigation due to selection rules. So, Brewster's or multipass geometry configuration should be used. Preliminary studies of intersubband absorption at Brewster's angle incidence showed the low level of absorption and presence of interference of *s*-polarized light on thin film. Therefore, the multipass geometry of experiment was chosen. Two opposite edges of the sample were ground away at a 45 degrees angle (see figure 1). It resulted in an increase of the optical path length and an improvement of the measurements accuracy.

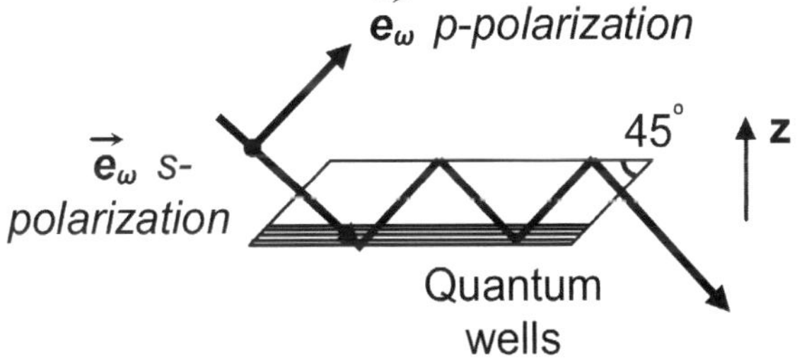

Figure 1. Multipass geometry scheme.

The samples were characterized by the investigation of the interband photoluminescence (PL) with a Horiba Jobin Yvon FHR640 monochromator. A CCD-camera Symphony II was used as a detector. A solid-state YAG laser with 532 nm radiation wavelength was used for pumping. The samples were placed in a liquid nitrogen bath cryostat. It allowed operating in the temperature range from 77 to 300 K.

The spectra of intersubband absorption were obtained with a vacuum Fourier transform spectrometer Bruker Vertex 80v paired with different photodetectors (MCT, pyroelectric, silicon bolometer). A globar was used as a source of infrared radiation. The sample was placed into a closed cycle cryostat Janis PT407RM with operational temperature range of 4–320 K and a temperature setting accuracy of 0.1 degrees.

3. Results and discussions

The interband photoluminescence spectra were measured in order to determine the real electron energy spectrum of the structure. The investigations were carried out at different temperatures and different optical pumping levels for the structure as-grown (curve 1, figure 2) and for the structure with a removed doped surface layer (curve 2, figure 2). It was found that the heavily doped surface GaAs layer significantly changes the PL spectrum. The obtained positions of the PL peaks of the structure without the top conductive layer are in a good agreement with the theoretical calculation of electron transition energies at room temperature: e1-hh1 = 1,486 eV and e2-hh2 = 1,525 eV.

The photoluminescence spectra of the structure without doped GaAs layer were then measured for various lattice temperatures (77-300 K) and different levels of optical pumping. Figure 3 shows the PL spectra for optical pump power about 550 mW. The photoluminescence spectrum shift to longer wavelengths with increasing temperature is associated with a decrease of the bandgap. Also, the temperature increase causes the changes of the carrier distribution function. It results in the increase of the electron concentration at the second electron level, therefore, transitions e2-hh2 become possible (see figures 2 and 3).

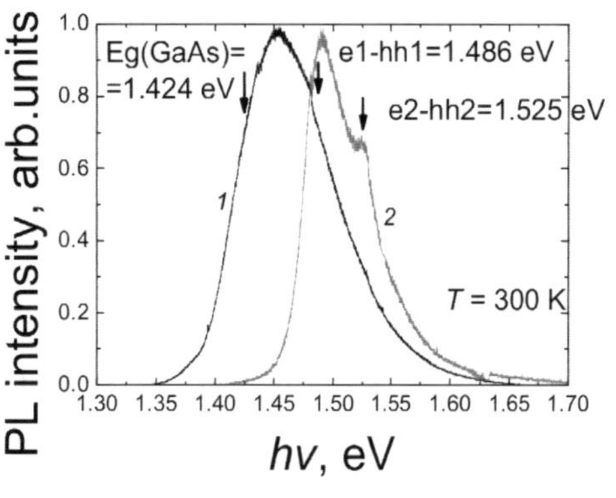

Figure 2. Normalized photoluminescence spectra of structures with tunnel-coupled GaAs/AlGaAs QWs at room temperature: structure with top doped layer (curve 1); structure without doped surface layer (curve 2). The arrows show the calculated energies of corresponding carrier transitions and the bandgap of GaAs. Optical pumping power was about 200 mW.

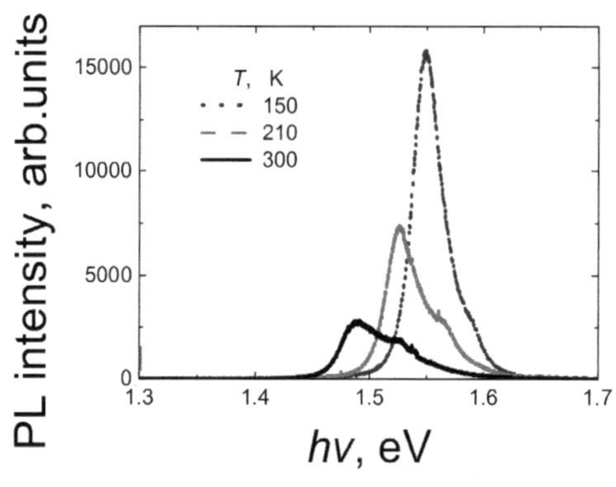

Figure 3. The photoluminescence spectra of tunnel-coupled QWs GaAs/AlGaAs without top doped GaAs layer at different temperatures. Spectral resolution was less than 1 meV.

The intersubband absorption spectra were studied for two light polarizations (*s* and *p*) in multipass geometry (see figure 1) for the temperature range of 4 to 300 K. According to the intersubband transitions selection rules, only *p*-polarized light (which has the polarization component along the structure growth axis) can be absorbed by QW layers. The typical transmission spectra for two light polarizations at 4 K is presented in figure 4. The absorption peak for *p*-polarized light (~130 meV) corresponds to the electron transitions from the ground subband e1 to the third subband e3. Its position is in a good agreement with the theoretical calculation of e1-e3 transition energies.

The spectrum of *p*-polarized light transmission (I_p) was normalized to the spectrum for *s*-polarized light (I_s) in order to estimate the intersubband absorption of quantum wells only (α_{QW}). According to the Beer–Lambert–Bouguer law,

$$\ln(\frac{I_p}{I_s}) = \ln(\exp((\alpha_p - \alpha_s)L)) = (\alpha_p - \alpha_s)L = \alpha_{QW}L\,, \tag{1}$$

where L is optical length. Figure 5 shows the intersubband absorption spectra at different temperatures.

The absorption peak at approximately 135 meV (see figure 5) corresponds to e1-e3 electron transitions. Absorption related to these transitions decreases with the temperature. Simultaneously temperature increase leads to an increase of the absorption intensity at the wavelength corresponding to e2-e3 transitions but we were not able to study light absorption in this spectral range. We associate variation of absorption with redistribution of electrons between electron levels e1 and e2. Slight broadening of the e1-e3 peak with increasing temperature can be attributed to the thermal broadening of the electron levels. Thus, the influence of the change of lattice temperature on the light absorption is similar to the influence of the vertical electric field.

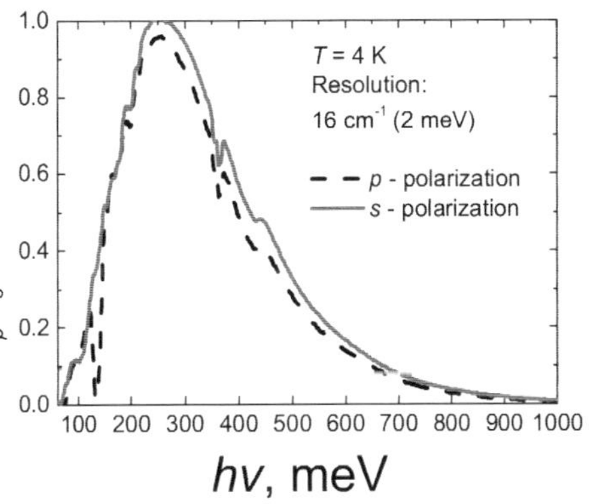

Figure 4. Transmission spectra of tunnel-coupled GaAs/AlGaAs QWs for p- and s-polarized light (dashed and solid line) at 4 K. Spectral resolution was about 2 meV.

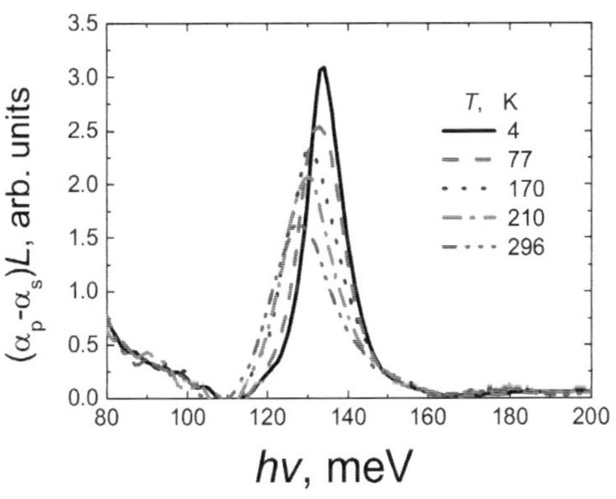

Figure 5. Intersubband absorption spectra, measured at different temperatures.

We also studied the spectra of intersubband absorption modulation under the conditions of transverse electric field application for p-polarized light in multipass geometry at liquid nitrogen temperature (figures 6 and 7). As expected, electro-optic modulation was not registered in the investigated spectral range for s- polarized light.

We have not found significant changes in absorption in the electric field of negative polarity (see figure 7). It can be explained by the fact that negative electric field does not change the relative position of the levels e1 and e2 (see inset in figure 7). In the positive electric field, a decrease of absorption associated with e1-e3 transitions was observed (see figure 6). This can be attributed to the transitions of the carriers from the ground QW level e1 to the second QW level e2 (see inset in figure 6). The spectra for the voltage value of 20 V and 25 V are almost identical. This saturation of modulation can be explained by the complete redistribution of carriers from level e1 to e2.

4. Conclusion

The optical properties of double tunnel-coupled GaAs/AlGaAs quantum wells in mid-infrared spectral range have been investigated under the conditions of applied vertical electric field. Intersubband absorption modulation was registered and attributed to the electrons redistribution between first two energy levels. The temperature modification of the intersubband absorption spectra was studied as well.

Acknowledgments

The authors are grateful to V.Yu. Panevin for the invaluable help in the experiments. This work was supported by the Russian Foundation for Basic Research (grants 14-02-00336, 14-02-31489, 13-02-12203, 12-02-01155), and the Ministry of Education and Science of the Russian Federation.

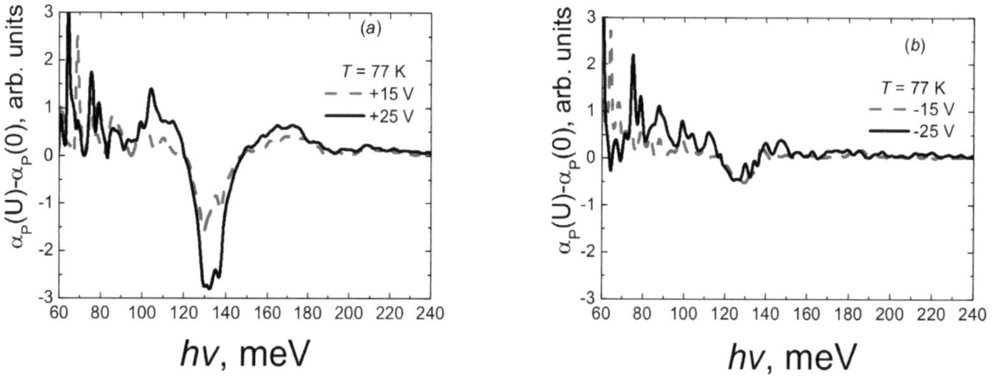

Figure 6. Modulation of intersubband absorption spectra under positive (*a*) and negative (*b*) transverse electric field.

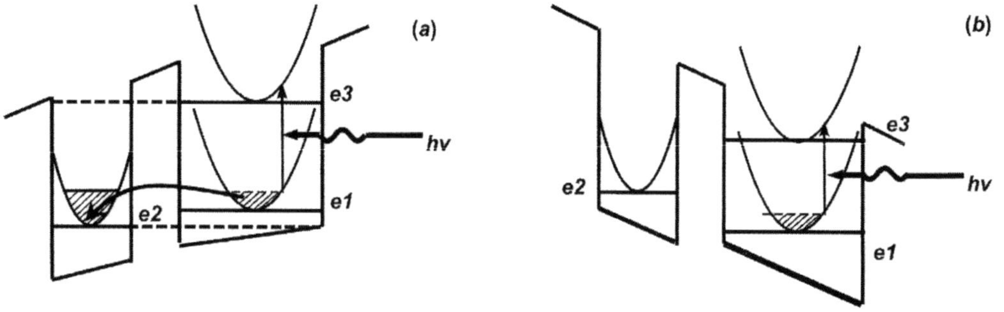

Figure 7. Redistribution of electrons in real space in the system of double tunnel-coupled QWs caused by positive (*a*) and negative (*b*) transverse electric field.

References

[1] Lurvi S, Gouzman M 2006 *International Journal of High Speed Electronics and Systems* **16**(2) 559–566
[2] Martini R, Bethea C, Capasso F, Gmachl C, Paiella R, Whittaker E A, Hwang H Y, Sivco D L, Baillargeon J N, Cho AY 2002 *Electronic letters* **38**(4) 181–183
[3] Sa'ar A, Kapon R 1997 *IEEE Journal of Quantum Electronics* **33**(9) 1517–1526
[4] Holm D A, Taylor H F 1989 *IEEE Journal of Quantum Electronics* **35**(1) 2266–2271
[5] Harwit A, Fernandez R, Eades W D, Kinell D K, Whitney R L 1996 *Superlattices and Microstructures* **19**(1) 39–45
[6] Dupont E, Delacourt D, Berger V, Vodjdani N, Papuchon M 1993 *Applied Physics Letters* **62**(16) 1907–1909
[7] Kapon R, Cohen N, Sa'ar A, Thierry-Mieg V, Planel R 1999 *Applied Physics Letters* **75**(11) 1583–1585
[8] Kuo D M-T, Chang Y-C 2000 *Journal of Applied Physics* **87**(6) 2936-2940
[9] Segev A, Sa'ar A, Oiknine-Schlesinger J, Ehrenfreund E 1996 *Superlattices and Microstructures* **19**(1) 47-54

Boron trichloride plasma treatment effect on ohmic contact resistance formed on GaN-based epitaxial structure

A A Kobelev[1], Yu V Barsukov[2], N A Andrianov[3] and A S Smirnov[1]

[1] St. Petersburg Polytechnic University, 29 Polytechnicheskaya str., St. Petersburg 195251, Russia
[2] Coddan Technologies LLC, 22 Beloostrovskaya str., St. Petersburg 197342, Russia
[3] Svetlana-Rost JSC, 27 Engels ave., St. Petersburg 194156, Russia

E-mail: kobelev_anton@mail.ru

Abstract. A BCl_3 plasma treatment effect on GaN surface and reduction of ohmic contact resistance in GaN/AlGaN epitaxial structure with a GaN cap layer has been studied. The BCl_3 plasma treatment was carried out by an inductively coupled plasma reactive ion etching (ICP RIE) system under conditions of low discharge power with zero external bias controlling power to prevent any recess etching of the GaN cap layer. The measured average contact resistances with and without plasma treatment were 1.8×10^{-5} and 1.05×10^{-5} $\Omega\cdot cm^2$, respectively. To study the BCl_3 plasma treatment effect on GaN surface and explain the increasing of contact resistance, the elemental composition of GaN surface was measured using x-ray photoelectron spectroscopy. It was found that a B_x–Cl_y layer covered the GaN surface after the BCl_3 plasma treatment. This can indicate that ion bombardment from plasma is inefficient to prevent deposition of BCl_x radicals on the GaN cap layer surface and subsequent formation of the B_x–Cl_y layer which is responsible for increase of ohmic contact resistance after the BCl_3 plasma treatment.

1. Introduction

The GaN-based high electron mobility transistors (HEMT) are promised to play a key role in the next-generation electronic devices such as mobile communication systems and pulsed radars. High breakdown fields, high saturation velocities and high thermal conductivities of GaN/AlGaN HEMTs result in its successful application at high powers, high frequencies and high temperatures [1]. To achieve excellent performance of these devices, it is important to obtain low-resistance ohmic contacts. Plasma enhanced etching processes in chlorine environments have been established to lower the contact resistance in GaN-based HEMT [2, 3]. However, the implementation of uniform ohmic contact is still difficult in case of large diameter wafers using this technology because of difficulty with provision of plasma enhanced etching uniformity.

Recently, Fujishima and co-authors [4] have reported a BCl_3 plasma treatment technique without any recess etching of the substrate implemented for low-resistance ohmic contact formation on GaN/AlGaN HEMTs with GaN cap layer using an Inductively Coupled Plasma Reactive Ion Etching (ICP-RIE) system. According to x-ray photoelectron spectroscopy and positron annihilation spectroscopy measurements they established that a surface oxide layer was efficiently removed from GaN cap layer and the vacancy-type defects were generated during the plasma treatment resulting in reduction of contact resistance [4]. Earlier other studies were also dedicated to the mechanism of GaN surface donor density increasing during the ion irradiation from plasma [5, 6].

In the present study BCl_3 plasma treatment of GaN cap layer without any recess etching have been performed using ICP-RIE system to obtain a lower contact resistance. To study its effect on GaN surface and subsequent ohmic contact resistance, the elemental composition of GaN surface after BCl_3 plasma treatment was measured using x-ray photoelectron spectroscopy (XPS).

2. Experimental setup

AlGaN/GaN structure with 2 nm GaN cap layer was deposited on SiC substrate using Molecular Beam Epitaxy (MBE) method. Photolithography was performed to pattern the ohmic contacts. Then, O_2 plasma ashing was performed to remove photoresist residuals, followed by a dip in pre-metallization solution of deluted HCl. After that one half of the wafer was exposed to BCl_3 plasma treatment in ICP-RIE system (Corial 210D) at power of 200 W under pressure of 10 mTorr and gas flow rate of 10 sccm for 1 minute of plasma treatment. External bias controlling power was set to zero because increasing of DC bias increases ion bombardment from plasma which results in recess etching of the sample. During the plasma treatment another half of the wafer was masked with a piece of quartz wafer and was not exposed to plasma. Then, ohmic contacts metallization Ti/Al/Ni/Au was performed followed by Rapid Thermal Anneal (RTA) at 830°C. Mesa isolation was formed by dry etching in BCl_3/Ar mixture in ICP-RIE system. Finally, contact resistance was measured by the Transmission Line Method (TLM). The width and length of the contacts were both 100 μm. The distance between contacts was of 5, 10, 20, 40, 200 μm as it is shown in figure 1.

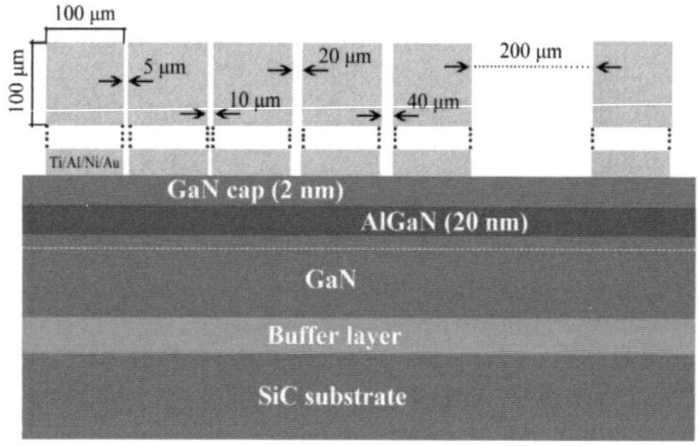

Figure 1. Schematic TLM structure of AlGaN/GaN epitaxial structure with 2 nm GaN cap layer and ohmic contact metallization.

3. Results and discussion

In the case of substrate exposed to BCl_3 plasma treatment for 1 minute the measured average ohmic contact resistance was 1.8×10^{-5} $\Omega\cdot cm^2$ which is 1.7 times higher than for the untreated one (1.05×10^{-5} $\Omega\cdot cm^2$). To study the BCl_3 plasma treatment effect on surface of GaN cap layer and increase of ohmic contact resistance, XPS analysis of GaN cap layer was carried out using a VG Escalab technique. The surface chemical composition of the substrate after plasma treatment step was characterized by monitoring the B 1s, C 1s, Cl 2p, Ga 3d, N 1s and O 1s peaks. The high atomic percentages of boron (29.7%), chlorine (35%) and carbon (20.6%) elements were observed at the surface after exposure to BCl_3 plasma. Summarized information on measured atomic percentage is presented in the table 1. Very low XPS signal coming from GaN substrate (gallium and nitrogen atomic elements) was obtained as it is shown from the measured survey spectrum in figure 2. High percentage of carbon at GaN surface is explained by exposure to ambient atmosphere and difficulties with its removal during the measurements.

Table 1. Atomic composition of GaN cap layer surface after BCl_3 plasma treatment obtained from XPS measurements.

Element peak	Binding energy peak (eV)	Atomic percentage (%)
Cl 2p	200.31	34.97
B 1s	190.39	29.67
C 1s	284.85	20.59
O 1s	532.71	8.71
Ga 3d	19.19	3.11
N 1s	402.58	1.74

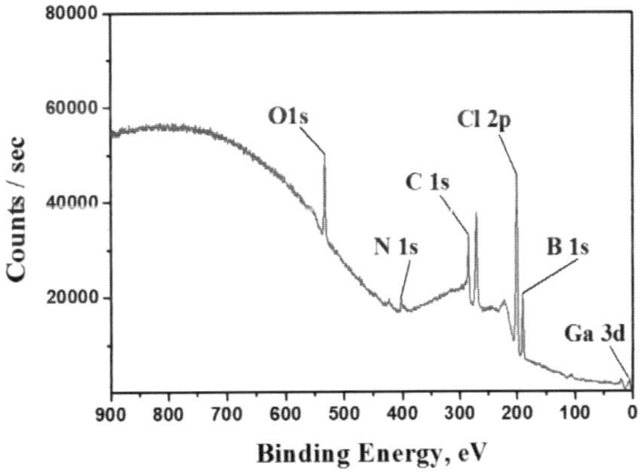

Figure 2. XPS survey spectrum measured after BCl_3 plasma treatment of GaN cap layer.

Two high peaks characterizing a B–Cl bond obtained from the analysis of GaN surface after exposure to BCl_3 plasma are also reported. The B 1s peak located at 190.2 eV is attributed to B bonded to Cl and Cl 2p one located at 199.7 eV is attributed to Cl bonded to B. Therefore, we can assume an existence of B_x–Cl_y layer on the GaN surface after BCl_3 plasma treatment step. The corresponding high-resolution spectra in the energy range of the B 1s and Cl 2p signal are shown in figure 3 and figure 4, respectively.

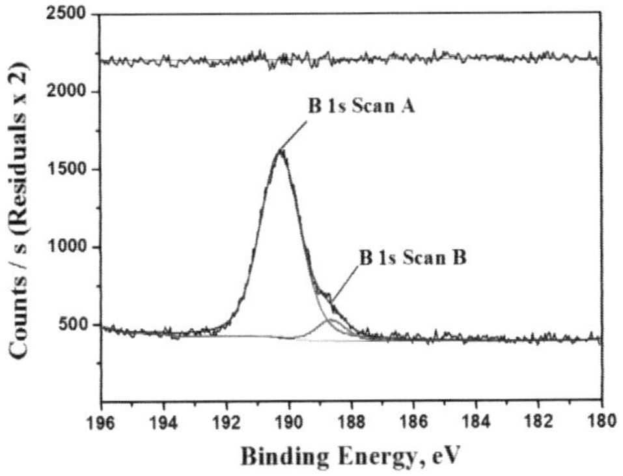

Figure 3. High-resolution B 1s spectrum obtained from XPS measurements of GaN surface after plasma treatment.

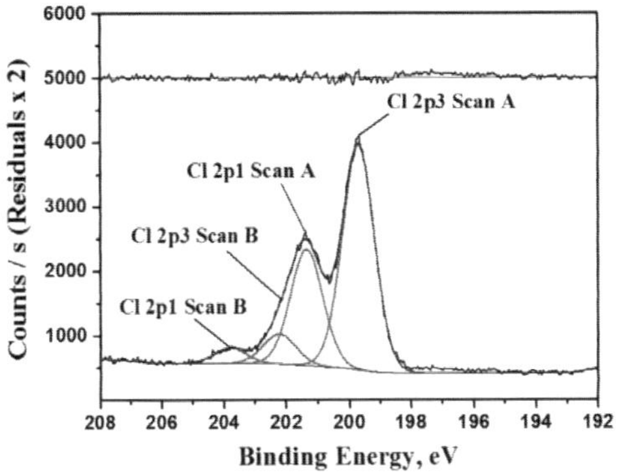

Figure 4. High-resolution Cl 2p spectrum obtained from XPS measurements of GaN surface after plasma treatment.

Another chlorine binding state was observed at binding energy of 201.4 eV and could be attributed to Cl–O bond because of possible oxygen incorporation from the quartz walls of the reactor during the plasma treatment [7].

It is known that exposure to BCl_3 plasma is a good recipe to remove Ga_xO_y layer formed after O_2 ashing step and exposure to ambient atmosphere [4, 8]. During the plasma processing BCl_x radicals generated by electron impact dissociation increase oxide layer etching by forming volatile B_xOCl_y and B_xO_y surface by-products which are removed from substrate by ion bombardment from plasma. On the other hand, if ion bombardment is weak due to low ions energy and flux, removal of the volatile by-products from the surface covered by thin oxide layer will be inefficient and the deposition of BCl_x radicals can lead to formation of B_x–Cl_y layer during the plasma processing. The analogous mechanism describing an impact of ion bombardment on a transition process from the deposition regime to the recess etching in case of HfO_2 and SiO_2 etching by BCl_3 plasma was reported by Sungauer and co-authors [7]. Therefore, in our case it is necessary to increase ion energy and flux by application of external bias controlling power to remove B_x–Cl_y layer and induce nitrogen vacancies during the BCl_3 plasma treatment step without any recess etching of the GaN cap layer. The measured average contact resistance in case of increased ion energy was 6.79×10^{-6} $\Omega\cdot cm^2$ which is 1.54 times lower than for the untreated one (1.05×10^{-5} $\Omega\cdot cm^2$).

4. Conclusion

In present work BCl_3 plasma treatment effect on surface of GaN cap layer of AlGaN/GaN epitaxial structure has been studied. Plasma treatment was performed in ICP-RIE system at low ICP power and zero external bias controlling power. It was found that exposure to BCl_3 plasma treatment resulted in increase of the ohmic contact resistance, in contrast to the results from the [4]. The measured average ohmic contact resistances after plasma treatment and without it were 1.8×10^{-5} and 1.05×10^{-5} $\Omega\cdot cm^2$, respectively. In order to study the BCl_3 plasma treatment effect on GaN cap layer surface and subsequent increase of contact resistance, the chemical composition of the GaN surface was characterized by monitoring the B 1s, C 1s, Cl 2p, Ga 3d, N 1s and O 1s peaks using XPS. The measurements have shown the high atomic percentage of boron (29.7%), chlorine (35%) and carbon (20.6%) elements and small signal from GaN substrate elements (Ga and N). C 1s peak characterizes the carbon contamination from ambient atmosphere. The obtained B 1s and Cl 2p peaks characterize the B–Cl bonds. This indicates that the B_x–Cl_y layer covers the GaN surface because of inefficient removal of the volatile by-products and subsequent deposition of BCl_x radicals on the GaN surface covered by thin oxide layer. The analogous mechanism was described previously for HfO_2 and SiO_2 etching in BCl_3 plasma [7]. The increase of ion energy during the BCl_3 plasma treatment by applying additional bias controlling power resulted in 1.54 times decrease of contact resistance which can be explained by removal of B_x–Cl_y layer and subsequent reduction of ohmic contact resistance.

References

[1] Pengelly R, Wood S, Milligan J, Sheppard S and Pribble W 2012 *IEEE Transactions on Microwave Theory and Techniques* **60** 1763–83
[2] Arulkumaran S, Vicknesh S, Ng G, Liu Z, Bryan M and Lee C 2010 *Electrochem. Solid-State Lett.* **13** H169
[3] Lee H.-S, Lee D and Palacios T 2011 *IEEE Electron Device Letters* **32** 623
[4] Fujishima T, Joglekar S, Piedra D, Lee H.-S, Zhang Yu, Uedono A and Palacios T 2013 *Appl. Phys. Lett.* **103** 083508
[5] Noor Mohammad S 2004 *J. Appl. Phys.* **95** 4856
[6] Nord J, Nordlund K and Keinonen J 2003 *Phys. Rev. B* **68** 184104
[7] Sungauer E, Pargon E, Mellhaoui X, Ramos R, Cunge G, Vallier L, Joubert O and Lill T 2007 *J. Vac. Sci. Technol.* B **25** 1640
[8] Buttari D *et al* 2003 *Appl. Phys. Lett.* **83** 4779

Experimental model of multi-junction solar cell: IV curve tailoring and photoresponce artifact simulation

S A Kozhukhovskaia[1,2], E D Filimonov[1,2] and M Z Shvarts[1]

[1] Ioffe Physical-Technical Institute, St. Petersburg, 194021, Russia
[2] BSTU, St. Petersburg, 190005, Russia

E-mail: svetlana07.87@mail.ru

Abstract. This paper presents an experimental model (maquette) of a multijunction solar cell, which allows one to study the IV characteristic construction process, the influence of individual p–n junctions on the resulting IV characteristic and the causes of artifacts in measuring the spectral dependencies of external quantum efficiency.

1. Introduction

State-of-the-art high-efficient multijunction solar cells (MJ SCs) consist of some single-junction solar cells stacked upon each other, so that each layer going from the top to the bottom has a smaller band gap than the previous one, and so it absorbs and converts the photons that have energies greater than the band gap of that layer and less than the band gap of the higher layer [1]. Thus, MJ SCs based on multiple materials with different band gaps span the solar spectrum. In order to optimize conversion efficiency of a MJ SC, the photovoltaic device should absorb as much of the sunlight as possible, and so band gaps should cover a wide range. Besides, band gaps of adjacent layers should differ by the smallest possible amount, because the amount of excess energy from light converted into heat is equal to the difference between the photon energy and the band gap of the absorbing material [2]. The efficient GaInP/GaAs/Ge triple-junction solar cells being currently in production are made of GaInP (1.9 eV), GaAs (1.4 eV) and Ge (0.7 eV) (figure 1).

In recent years, several analytical methods were proposed for simulating the IV characteristics of MJ SC, whereas experimental models of MJ SCs were not elaborated and almost not mentioned in the literature [3–5]. However, the experimental simulation of the MJ SC IV characteristic is an important tool for investigating the performance of solar cells at variable irradiance, temperature and solar spectrum.

For a MJ SC with two electrical terminals, it is necessary to know the characteristics of the p-n junction of each subcell individually. Existing research methods make it possible to be performed. In particular, in recording the spectral characteristics and allocating a signal from the test subcell and excluding the influence of other subcells, one can affect selectively the individual subcells by varying the spectral composition of the incident radiation in recording the IV characteristics.

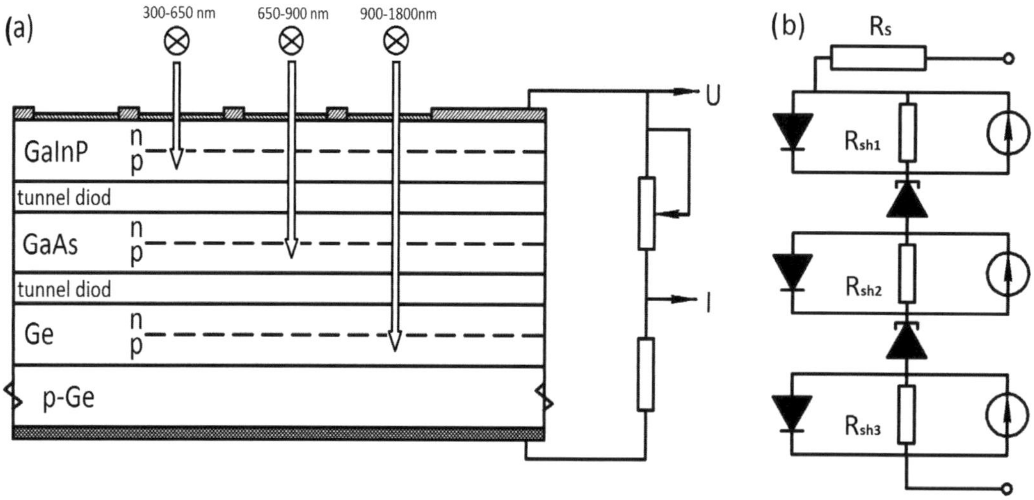

Figure 1. Structure of a triple-junction solar cell (a) and its equivalent circuit (b).

2. Experimental details

For detailed study and understanding of MJ SC IV curves formation processes and the influence of individual p-n junctions on the resulting IV characteristic and on the accuracy in obtaining the spectral dependence of photosensitivity, the specialized maquette is required to be designed and created.

In such a MJ SC maquette, each element allows simulating physically and photo-electrically the behaviour of individual subcells, tunnel diodes, variable series and shunt resistances and also diodes forming the reverse branch of IV curves.

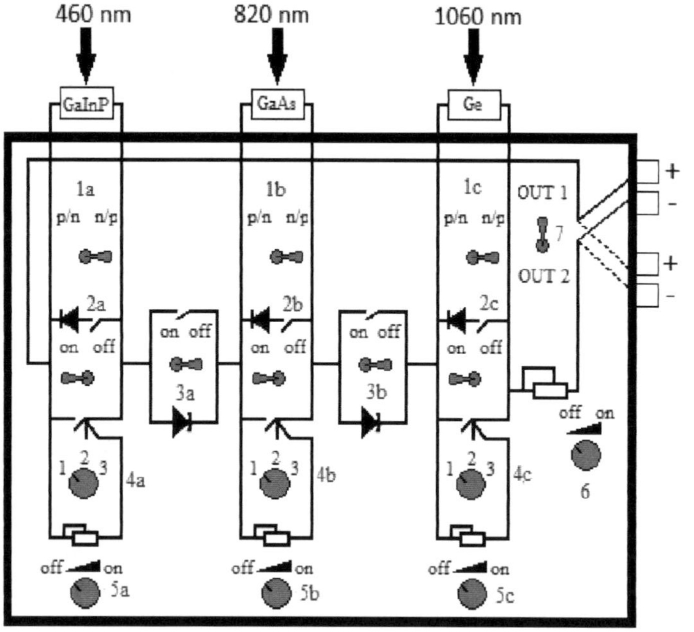

Figure 2. Circuit of the triple-junction SC experimental model. Set of switches allows one to: 1a,b,c –change polarity of individual subcells; 2a,b,c – connect diodes for forming reverse branch of subcell IV curves; 3a,b – connect tunnel diodes; 4a,b,c – connect variable shunt resistors; 5a,b,c: pos.1 – subcells are in short circuit conditions, pos.2 – no shunt resistor in the circuit, pos.3 – shunt resistors are in the circuit; 6 – connect variable series resistor; 7 – connect the maquette either to the IV characteristic recording system (OUT1) or to the IV tracer with an oscilloscope (OUT2).

The maquette consists of three identical electrical circuits, each of which is an analog of a subcell and includes a single-junction SC, a diode connected to it in parallel and a shunt variable resistance. Subcells are combined in an electrical circuit via tunnel diodes connected in series. The whole circuit is connected to an IV characteristic loading setup via a variable resistor, which plays a role of the series resistance. Depending on the task of forming the IV curves, the maquette circuit allows one to change the operation modes of subcells by the illumination level and to vary their electrical parameters (to change SC polarity, to vary the shunt resistor, to connect diodes, etc.) to the complete exclusion of any of them from the series circuit. For analyzing the shape-formation of IV curves, two output setups are provided: one is for connecting to an analog IV tracer (oscilloscope) and another one is for digital IV curve recording.

The practical maquette includes single-junction SCs of the following configurations: GaInP, GaAs with a GaInP filter, Ge with a GaAs filter with the spectral sensitivity in the wavelength ranges of 300–650, 650–900 and 900–1800 nm, respectively. Selective impact on the individual solar cells is provided by a light source based on two high-brightness 460 and 820 nm LEDs and a 1060 nm laser diode.

The elaborated maquette allows tailoring the MJ SC IV curves, investigating the effect of individual subcells on the resulting characteristic, determining the IV curve behavior dynamics in the modes of the tunnel diode peak current limitation and at increased values of the shunt series resistance. It is easy to manipulate with the reverse branch of IV curves by counter connected diodes. It is also possible to perform mismatch of the subcells′ photocurrents in a wide range of generated currents.

3. Results and discussion

For a practical solar cell, typical is the presence of a series resistance, a resistance of each p-n junction, as well as the shunt resistance that reflects possible surface and bulk current leakages. Series and shunt resistances cause losses in SCs, which lead to the fill factor and the open circuit voltage drop. As an example, figure 3 shows a simulated multi-step IV characteristics formed as composite of individual subcells curves. Changing shunt resistance of one of the subcell affects the shape of a MJ SC IV curve as a whole, and also can affect another current AC amplitude ($I_{\sim}$) measured at a certain voltage bias (U_{bias}).

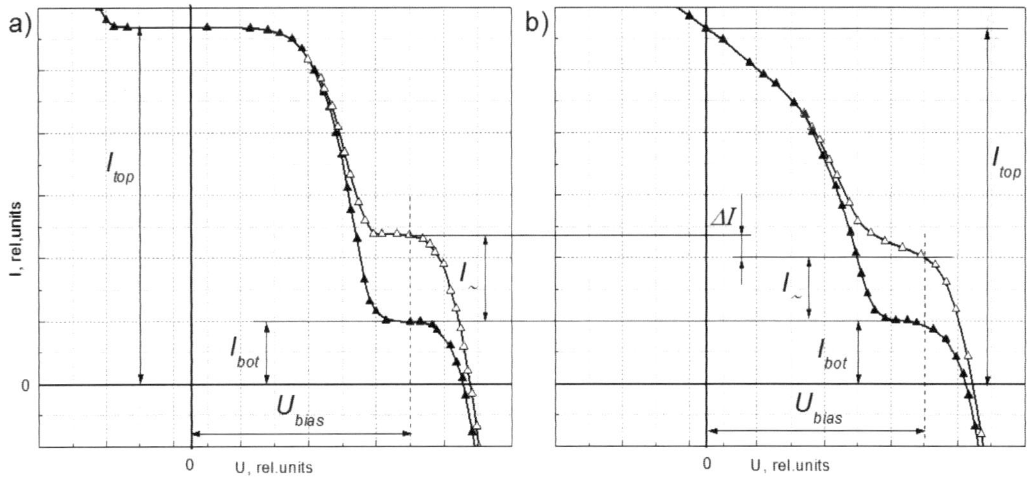

Figure 3. IV curve of a MJ SC in conditions of current limitation by the bottom narrowband subcell photocurrent: without (a) and pronounced (b) top subcell shunt resistance: I_{top} and I_{bot} – photocurrents generated in top and bottom subcells at steady-state illumination, correspondingly; $I_{\sim}$ – photocurrent generated in bottom subcell at chopped (AC) illumination; ΔI – difference between bottom subcell photocurrent amplitudes resulting from the impact of the shunt resistance.

In studying the photoresponce or spectral dependence of the external quantum efficiency (EQE) of individual subcells, the functional capability of the maquette allows considering individual subcells and affecting them with the aim to achieve the conditions for the mutual impact of their characteristics on the accuracy of the EQE determination and the causes of the "anomalous" photosensitivity outside the operating wavelength range typical for a tested subcell. This effect is called EQE artifact, and it appears if the subcell has a low shunt resistance or a low reverse breakdown voltage [6–8]. The main aspect of the artifact measurement is that the response of a passive subcell is simultaneously measured with the signal from the bottom subcell. The magnitude of the photoresponce artifact depends on the shunt resistance of the subcell under test and the difference between the bias currents of other subcells. The magnitude of the parasitic signal rises with lowering the shunt resistance and/or could be expressed in a small difference between the currents produced by light biasing.

4. Conclusion

The maquette presented allows investigating the influence of the shunt resistance on the tailoring of IV curves of MJ SCs and also on the procedure and measurement accuracy in terms of physics rather than mathematics, and gives an opportunity for visualizing the contribution to PV parameters being measured. The results obtained with the use of the present maquette are well agreed with theoretical and experimental data for monolithic MJ SCs. The experimental maquette is adapted to carry out scientific research in the field of photovoltaic sunlight converters and to study in door for "Optoelectronics and Nanophotonics" education topics.

Acknowledgments

This work was supported by the Russian Science Foundation (Grant № 14-29-00178).

References

[1] Burnett B 2002 *The basic physics and design of III-V multijunction solar cells* (Golden, CO: National Renewable Energy Laboratory)

[2] Zweibel K 1984 *Basic photovoltaic principles and methods* (New York: Van Nostrand Reinhold)

[3] Nishioka K, Takamoto T, Agui T, Kaneiwa M, Uraoka Y and Fuyuki T 2006 Annual output estimation of concentrator photovoltaic systems using highefficiency InGaP/InGaAs/Ge triple-junction solar cells based on experimental solar cell's characteristics and field-test meteorological data *Solar Energy Materials and Solar Cells* **90** 57–67

[4] Adelhelm R and Bücher K 1998 Performance and parameter analysis of tandem solar cells using measurements at multiple spectral conditions *Solar Energy Materials and Solar Cells* **50** 185–195

[5] Emelyanov V M, Kalyuzhnyy N A, Mintairov M A, Mintairov S A, Shvarts M Z and Lantratov V M 2010 Distributed resistance effects simulation in concentrator MJ SCs using 3D-network model *Proc. of the 5th World Conf. on Photovoltaic Energy Conversion, Valencia, Spain* pp. 406–411

[6] Meusel M, Baur C, Letay G, Bett A W, Warta W and Fernandez E 2003 Spectral response measurements of monolithic GaInP/Ga(In)As/Ge triple-junction solar cells: Measurements artifacts and their explanation *Progress in Photovoltaics: Research and Applications* **11** 499–514

[7] Lim S H, Li J J, Steenbergen E H and Zhang Y H 2011 Luminescence coupling effects on multi-junction solar cell external quantum efficiency measurements *Progress in Photovoltaics: Research and Applications* **10** 102

[8] Siefer G, Baur C and Bett A W 2010 External quantum efficiency measurement of germanium subcells: Measurement artifacts and correction procedures *Proc. of the 35th IEEE Photovoltaic Specialists Conf., Honolulu, Hawaii* pp. 704–707

Diffusion-drift model of injection lasers with double heterostructure

I Pisarenko, E Ryndin and M Denisenko
College of Electronics and Electronic Equipment Engineering, Southern Federal University, 44 Nekrasovsky st., GSP-17A, Taganrog 347928, Russia

E-mail: ivan123tgn@yandex.ru

Abstract. This paper introduces the diffusion-drift model for numerical analysis of dynamical processes in injection lasers. The model takes into account the laser's structural features, the mechanisms of spontaneous and stimulated recombination. The numerical simulation of injection lasers with a double heterostructure has been made. The simulation results allowed to determine the validity limits for the proposed model.

1. Introduction

The paper is devoted to the problem of the simulation of physical processes in semiconductor laser structures. This problem is one of the urgent problems of modern semiconductor optoelectronics. It was considered in [1], [2], [3].

Currently the dynamics of processes in injection lasers is described by the kinetic equations which connect the concentration of charge carriers with the density of pumping current and the density of photons in the active region of the laser and are formulated as below [4], [5], [6]:

$$\frac{dn}{dt} = \frac{j}{eD} - \frac{n}{\tau_S} - v_g n_{ph} g(n, n_{ph}); \tag{1}$$

$$\frac{dn_{ph}}{dt} = -\frac{n_{ph}}{\tau_f} + \beta \frac{n}{\tau_S} + v_g n_{ph} g(n, n_{ph}), \tag{2}$$

where n is the concentration of mobile charge carriers in the active region of the laser; n_{ph} is the concentration of photons; j is the density of the pumping current; t is the time; e is the electron charge; D is the characteristic size of the active region of the laser; $g(n, n_{ph})$ is the optical gain coefficient; β is the fraction of spontaneous emission that falls in the laser mode; τ_S is the time of spontaneous radiative recombination; τ_f is the photon lifetime in the active region of the laser; v_g is the photon speed in the active region of the laser.

The kinetic equations represent a system of ordinary differential equations and do not take into account the following factors that can be important for analyzing the characteristics of the laser structures [7], [8]:

- the difference in the distribution of electron and hole concentrations by coordinates in the active region of the laser (the kinetic equations are derived on the assumption that electrons and holes concentrations are equal within the active region of the laser);
- uneven spatial distributions of charge carrier and photon concentrations in the active region;

- the features of spatial distribution of current density;
- the influence of peripheral regions on the laser characteristics;
- the impossibility of the transient analysis in the injection laser in the fixed voltage on contacts mode (only the fixed injection current mode was researched).

The aim of this work was to develop the physical and topological model of semiconductor laser structures which would consider the above factors.

2. Mathematical methods

This model was derived on the basis of the semiconductor fundamental equations system in diffusion-drift approximation [9] and the kinetic equations for lasers [4], [5], [6], which can be written as follows:

$$\nabla(\varepsilon\cdot\nabla\varphi)=\frac{e}{\varepsilon_0}(n-p-N); \tag{3}$$

$$\frac{\partial n}{\partial t}=\nabla\left[\mu_n\left(-n\nabla(\varphi+V_n)+\varphi_T\nabla n\right)\right]-\frac{\sqrt{np-n_i^2}}{\tau_S}-v_g g\cdot n_{ph}; \tag{4}$$

$$\frac{\partial p}{\partial t}=\nabla\left[\mu_p\left(p\nabla(\varphi-V_p)+\varphi_T\nabla p\right)\right]-\frac{\sqrt{np-n_i^2}}{\tau_S}-v_g g\cdot n_{ph}; \tag{5}$$

$$\frac{dn_{ph}}{dt}=-\frac{n_{ph}}{\tau_f}+\frac{\beta\sqrt{np-n_i^2}}{\tau_S}+v_g g\cdot n_{ph}; \tag{6}$$

$$g=\begin{cases} g(n,p,n_{ph})>0, & E_{Fn}-E_{Fp}\geq E_C-E_V;\\ 0, & E_{Fn}-E_{Fp}<E_C-E_V,\end{cases} \tag{7}$$

where n, p are electron and hole concentrations; μ_n, μ_p are electron and hole mobilities; N is the effective impurity concentration; n_i is the intrinsic concentration; φ is the electrostatic potential; φ_T is the temperature potential; V_n, V_p are heterostructure potentials in the conduction and valence bands; ε is the dielectric constant of a semiconductor; ε_0 is the dielectric constant of vacuum; g is the optical gain coefficient; E_{Fn}, E_{Fp} are electron and hole imrefs; E_C, E_V are levels of the bottom of the conduction band and the top of the valence band.

The boundary conditions for the fixed voltage on contacts which were used for solving the system of equations (3) – (7) can be written as follows:

- on the ohmic contacts:

$$\varphi=\varphi_T\ln\frac{\frac{N}{2}+\sqrt{\left(\frac{N}{2}\right)^2+n_i^2}}{n_i^2}+U(t); \tag{8}$$

$$n=\frac{N}{2}+\sqrt{\left(\frac{N}{2}\right)^2+n_i^2}; \tag{9}$$

$$p=-\frac{N}{2}+\sqrt{\left(\frac{N}{2}\right)^2+n_i^2}; \tag{10}$$

$$n_{ph}=0, \tag{11}$$

where U is the voltage on the contact;

- on Schottky's contact:

$$\varphi = \varphi_T \ln \frac{\frac{N}{2} + \sqrt{\left(\frac{N}{2}\right)^2 + n_i^2}}{n_i^2} + U(t) - \varphi_0 ; \tag{12}$$

$$n = \left(\frac{N}{2} + \sqrt{\left(\frac{N}{2}\right)^2 + n_i^2} \right) \exp\left(-\frac{\varphi_0}{\varphi_T} \right); \tag{13}$$

$$p = \left(-\frac{N}{2} + \sqrt{\left(\frac{N}{2}\right)^2 + n_i^2} \right) \exp\left(\frac{\varphi_0}{\varphi_T} \right); \tag{14}$$

$$n_{ph} = 0 , \tag{15}$$

where φ_0 is the contact-potential difference of Schottky's junction;

- on the boundaries which are free from contacts:

$$\frac{\partial \varphi}{\partial \xi} = 0 ; \tag{16}$$

$$\frac{\partial n}{\partial \xi} = 0 ; \tag{17}$$

$$\frac{\partial p}{\partial \xi} = 0 ; \tag{18}$$

$$\frac{\partial n_{ph}}{\partial \xi} = 0 , \tag{19}$$

where ξ is the surface normal.

The boundary conditions for the fixed current density are similar to (9) – (19), the equation (8) is replaced by the following expression:

$$\frac{\partial \varphi}{\partial \xi} = \frac{\frac{j}{e} + \mu_n \left(\varphi_T \frac{\partial n}{\partial \xi} - n \frac{\partial V_n}{\partial \xi} \right) - \mu_p \left(\varphi_T \frac{\partial p}{\partial \xi} - p \frac{\partial V_p}{\partial \xi} \right)}{\mu_n n + \mu_p p} . \tag{20}$$

The proposed model does not consider the dimensional quantization of the energy spectrum of the charge carriers. Therefore it cannot be used for the simulation of laser structures with narrow quantum wells.

The photon lifetime in the active region of the laser τ_f is determined by the optical loss and the parameters of the resonator mirrors.

The optical gain coefficient g is considered to be one of the key parameters of injection lasers. The optical gain coefficient dependence on the charge carrier and photon concentrations in the active region of the laser is traditionally defined by the models introduced in [5], [6]. In the proposed system of equations (3) – (7) the optical gain coefficient value $g(n, p, n_{ph})$ is defined similarly by the expressions given in [5], [6]. In these expressions the generalized concentration of the charge carriers that implies the equality $n = p$ in the active region of the laser, is replaced by the expression $\sqrt{np - n_i^2}$ for taking into account the influence of electron, hole and photon concentrations. The expression being considered determines the degree of the electron and hole concentrations deviation from the equilibrium values more accurately.

In contrast to the kinetic equations, the system of equations (3) – (7) allows to investigate the changes of spatial distributions of electron, hole and photon concentrations in the active and peripheral regions of the laser. Thus, transient processes analysis is possible at a set pumping current or at a specified voltage on contacts in accordance with the boundary conditions used. It is important to use the heterostructure potential and profiles of dopant spatial distributions as a data source for the equations (3) – (7) in view of the injection lasers structures huge variety. It allows obtaining the results on the laser numerical simulation taking into account the various structural and topological features thereby expanding the scope of the proposed model validity. The numerical solution of the equation system (3) – (7) was performed using the finite difference method. The solution of the appropriate stationary problem was performed using the Newton's method in order to obtain the initial condition [9]. The system of equations used for the calculation of the initial conditions can be written as follows:

$$\nabla(\varepsilon \cdot \nabla\varphi) = \frac{e}{\varepsilon_0}(n - p - N); \tag{21}$$

$$\nabla\left[\mu_n\left(-n\nabla(\varphi + V_n) + \varphi_T \nabla n\right)\right] = \frac{\sqrt{np - n_i^2}}{\tau_S} + v_g g \cdot n_{ph}; \tag{22}$$

$$\nabla\left[\mu_p\left(p\nabla(\varphi - V_p) + \varphi_T \nabla p\right)\right] = \frac{\sqrt{np - n_i^2}}{\tau_S} + v_g g \cdot n_{ph}; \tag{23}$$

$$n_{ph} = \frac{\beta\sqrt{np - n_i^2}}{\tau_S\left(\dfrac{1}{\tau_f} - v_g g\right)}; \tag{24}$$

$$g = \begin{cases} g(n, p, n_{ph}) > 0, & E_{Fn} - E_{Fp} \geq E_C - E_V; \\ 0, & E_{Fn} - E_{Fp} < E_C - E_V. \end{cases} \tag{25}$$

3. Simulation results and discussion

Figures 1 – 3 show the results of the simulation of the double heterostructure laser n^+-$Al_{0,3}Ga_{0,7}As$/i-$GaAs$/p^+-$Al_{0,3}Ga_{0,7}As$ with the active region width of 50 nm, the photon lifetime in the resonator is $\tau_f = 3$ ps, the time of spontaneous radiative recombination is $\tau_S = 4$ ns, the proportion of spontaneous emission getting in the laser mode is $\beta = 10^{-4}$ for the pulse of the pumping current density with the duration of $\tau_P = 100$ ps and the amplitude of $j = 10^5$ A/cm^2.

The results shown in figure 1(a) were obtained by solving the kinetic equations. Figures 1(b–d) and figure 2 show the results obtained using the proposed model. The time dependences of the electron, hole and photon concentrations are shown for various cross-sections of the active region of the double heterostructure laser in figure 1(b–d).

Figure 2 shows that electron and hole concentrations in the active region of the laser are distributed irregularly. It explains the irregularity of the photon concentration distribution within the active region, which is presented in figure 3. That is the reason why, as it is seen in figure 1, the quantitative discrepancy of results obtained by solving the kinetic equations (figure 1(a)) and the system of equations (3) – (7) (figure 1(b–d)) for individual cross-sections of the laser active region is more than 50 %.

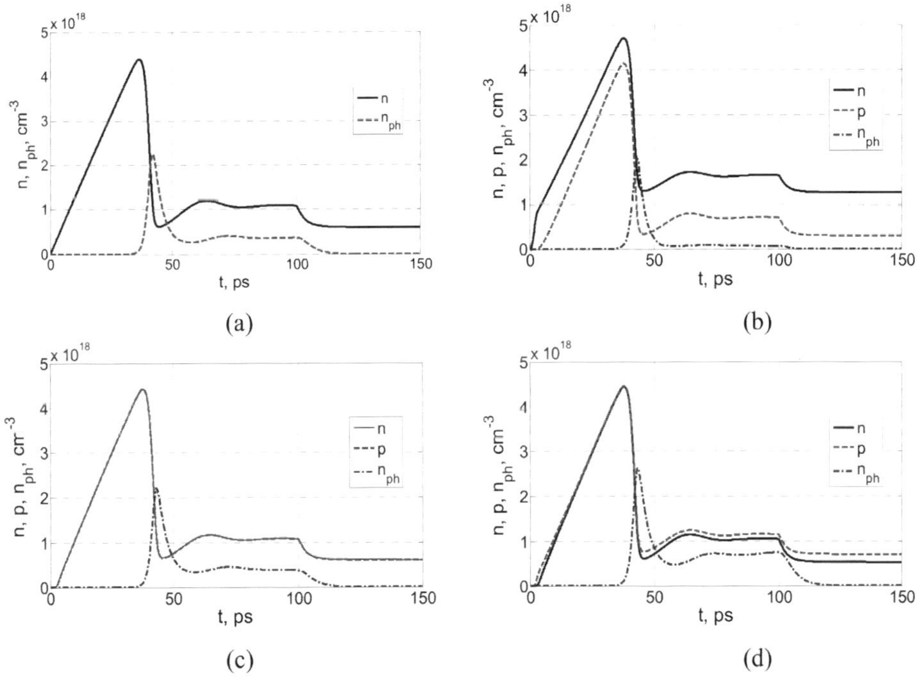

Figure 1. The results of the numerical simulation of the transient process in the double heterostructure laser obtained by solving the kinetic equations (a) and the equations of the proposed model for the central (c) and peripheral (b, d) cross-sections of the active region

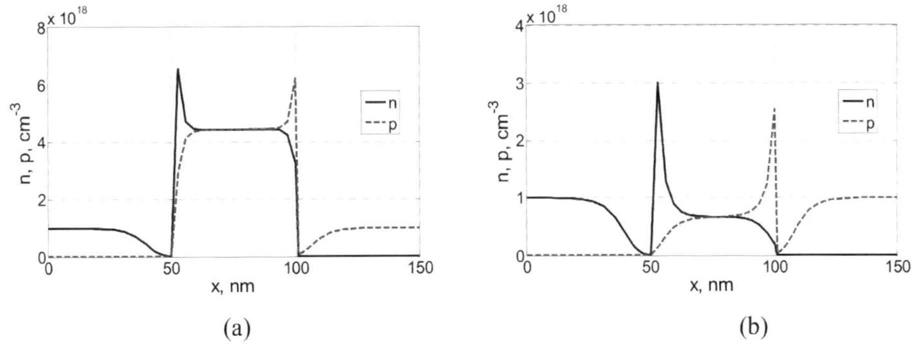

Figure 2. The electron and hole concentrations in the double heterostructure laser in the moments of reaching the first maximum (a) and the first minimum (b) in the central cross-sections of the active region

The kinetic equations for the laser were formulated under the following assumption: when the density of the pumping current is above the threshold value, the optical gain coefficient g undergoes periodic changes only over time in the course of the transient process. The numerical simulation results obtained using the proposed model (3) – (7) confirm the possibility of periodic changes of the laser optical gain coefficient values both in time and by coordinate. It allows analyzing the transient processes in the laser structures taking into account this effect.

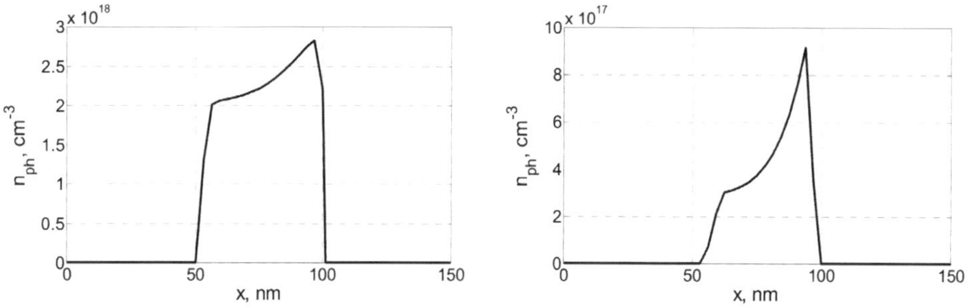

(a) (b)

Figure 3. The photon density in the moments of reaching the first maximum (a) and the current pulse rear front (b)

4. Conclusion

The comparative analysis of the obtained results allows to conclude that the kinetic equations with the accepted assumptions can be used for the simulation of transient processes in injection lasers if the problem doesn't require the consideration of non-uniformity of the spatial distributions of electron, hole and photon concentrations in the active and peripheral regions, non-uniformity of the laser optical gain coefficient by coordinates, characteristics of laser structure, its band diagram, spatial distribution of current density.

If all these factors (or some of them) have to be considered according to the problem of simulation or if it is needed to explore the transient processes in the laser for a set voltage pulse (not a current pulse) on the external contacts, it is advisable to use more accurate models, in particular, the equation system considered in this research.

The proposed approach allows to ensure the adequacy of the results of the laser structures simulation and to improve the accuracy of the calculation of the laser's characteristics.

Acknowledgements

Work is supported by the Russian Foundation for Basic Research (grants 13-07-00274, 14-07-31234) and the Ministry of Education and Science of the Russian Federation (project 213.01–11/2014–12).

References

[1] Sahli S, Lescure M, Boucher J 1994 *Optical and Quantum Electronics* **26** 97–107
[2] Wright P D 1987 *J. Appl. Phys.* **61** 1720-4
[3] Gao J 2011 *Optoelectronic Integrated Circuit Design and Device Modeling* (Beijing: Higher Education press)
[4] Ozyazici M 2004 *J. Optoelectr. Adv. Mater.* **6** 1243–53
[5] Lim D, Cho H, Sung H, Yi J and John Y 2009 *J. Opt. Soc. of Korea* **13** 386–91
[6] Zarifkar A, Ansari L and Moravvej-Farshi M 2009 *Fiber and Integrated Optics* **28** 249–67
[7] Konoplev B, Ryndin E and Denisenko M 2013 *Tech. Phys. Lett.* **39** 986–9
[8] Ryndin E and Denisenko M 2013 *Russian Microelectronics* **42** 360–2
[9] Abramov I 2006 *Nano- and microsystem technology* **9** 26–36

Relaxation phenomena in a naturally disordered Pb_3O_4 semiconductor

M P Sevryugina, N S Pshchelko and Ya S Kadi

National University of Mineral Resources «The University of Mines», Vasilevsky Island, 21st line, 2, St. Petersburg 199106, Russia

E-mail: marysevrugina@yandex.ru

Abstract. Charge transfer in Pb_3O_4 structures has been investigated. Electric current dependences on time at temperatures of 300–370 K and at the dc electric field strength in the range of $2\cdot10^5$–$9\cdot10^5$ V/m were measured. Flowing of relaxation polarizing current is shown to result in charge accumulation in the sample surface area. Experimental regularities coordinate with provisions of the relay mechanism of transfer of a charge with the participation of deep local levels. Values of the charge transfer parameters were determined.

1. Introduction

Among the components of the lead oxide series, Pb_3O_4 is distinguished by the stability of its phase composition in thin layers, photoelectric sensitivity in the visible spectral range, and pyroelectric and ferroelectric properties [1, 2]. Its high resistivity combined with a high density of deep localized states enables the use of the oxide Pb_3O_4 in electret structures. The defect and impurity states in high resistance Pb_3O_4 are responsible for the energy spectrum with a complex distribution of the energy levels of localized electron states in the band gap, which determines the charge state of the structure.

Polarization and charging processes determine the operation of most semiconductor devices, controlling the conditions of creation of the potential profile on the surface of photosensitive layers, as well as its variation under optical excitation. To establish the charge transport mechanisms and the nature of the local centers in high-resistivity photoconductors, experimental study and theoretical analysis of the mechanisms of polarization in electric fields of different intensities appeared to be effective.

Charge relaxation in semiconductor bulk can be assumed as one of the factors that determine the stability of characteristics of electronic elements. Relaxation phenomena in a high-resistivity structure are associated with the manifestation of any form of disorder since the possibility of revelation of relaxing elements in an ideally ordered system is reduced to a minimum. The investigation of temporal dependence of the currents of isothermal relaxation makes it possible to acquire information on the population kinetics of capture centers arranged nonuniformly over the semiconductor thickness as well as on the capture processes in near-electrode regions [3]. In the following, the results of the study of the polarization relaxation in photoconductive Pb_3O_4 layers are presented.

2. Experimental

Kinetic dependences of polarization currents were measured for the Pb_3O_4 layers with the binder—cyanoethyl ester of polyvinyl alcohol—deposited on a glass substrate with the ITO conducting film based on tin-doped indium oxides. A layer of current-conducting paste (aquadag) with the area of 1 cm2 and 50 μm thick was deposited on the surface of the material under study as the upper electrode. Currents of isothermal polarization were detected using a Keithley 6517A electrometric amplifier. The electric field strength was varied in the range of $E = 2\cdot10^5$–$9\cdot10^5$ V/m. The sample was heated using an electric furnace mounted into a holder. The furnace was dc-supplied from a stabilized current source. The heating rate in the limits of 3–5 K/min could be fixed in the range of $T = 300$–370 K.

3. Results and discussion

Figure 1 shows the results of the study of the kinetics of isothermal polarization currents in Pb_3O_4 layers measured for different electric fields. The shape of the polarization relaxation curves for Pb_3O_4 indicates that there are two phases of the process (fast and slow) corresponding to time intervals $\Delta t_1 =$ 1–30 s and $\Delta t_2 =$ 30–200 s. At the initial stage after switching the voltage on, rapid current decay is observed in polycrystalline Pb_3O_4 layers, obeying the empirical Curie–von Schweidler relation $I \propto t^{-n}$ ($n = 0.8$). A slow polarization kinetics component corresponds to the exponential dependence. As the voltage increases, the temporal range of the initial portion of the kinetics of the polarization current decreases, while the decay rate of function $I(t)$ increases.

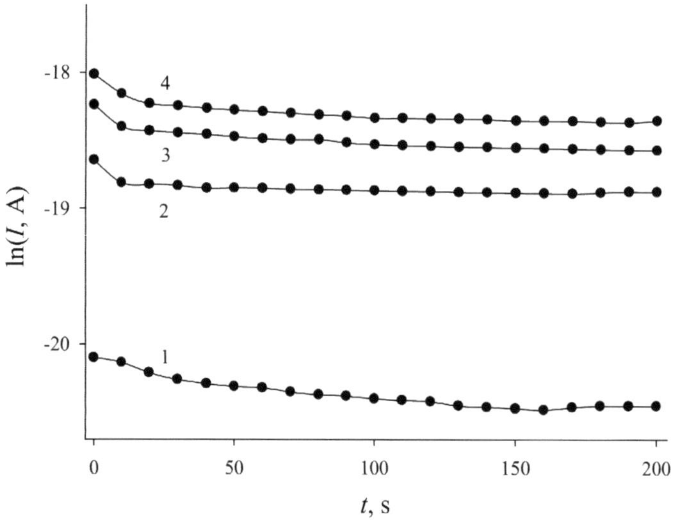

Figure 1. Kinetics of the polarization current measured for Pb_3O_4 layers at electric fields E: 1 – $2\cdot10^5$, 2 – $4\cdot10^5$, 3 – $6\cdot10^5$, 4 – $8\cdot10^5$ V/m. $T = 300$ K.

The data on the kinetics of the isothermal relaxation currents was interpreted within a model according to which the transfer of an electric charge injected into a crystal occurs through the relay-race mechanism [4, 5]. The charge transfer from the contact into the sample occurs through the mechanism of hopping conduction over the trap centers, which are located in the band gap and have deep trapping levels of free charge carriers. This process is also accompanied by the formation of an energy barrier at the boundary with the anode, which hinders the transfer of electrons (injected from the cathode) to the anode. The energy barrier at the boundary with the anode arises from the energy difference between the electron affinity for the local center and the work function of the anode metal. Thus, the electric current is limited by both the space charge in the bulk of the semiconductor and the energy barrier at the boundary with the anode, which leads to a redistribution of the voltage across the crystal and decrease of electric current passing through the crystal.

Long-time relaxation processes established for the structures under study on the basis of polycrystalline Pb_3O_4 layers can also be accompanied by the formation of depletion regions at the grain–intergrain layer boundary. A nonuniform potential profile of the boundary regions in the polycrystalline structure studied is the same as that created by a quasi-continuous energy spectrum of the traps.

The current relaxation decay observed for the structure is accompanied by the accumulation of charge whose value can be determined from the area lying under the curve of the time dependence of the current. In figure 2, the values of this charge are plotted as functions of the applied electric field. We can notice that in low electric fields, the accumulated charge increases to a first approximation by the quadratic law, while with the further increase, the polarization varies according to linear dependence.

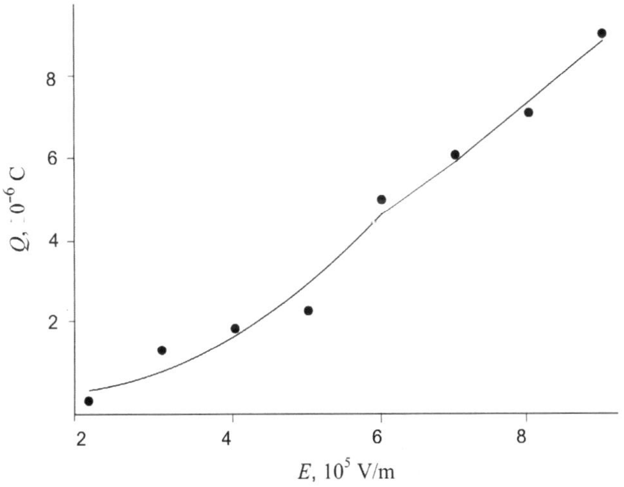

Fig. 2. Charge accumulated in the Pb_3O_4 layer and measured as a function of the electric field *E*. $T = 300$ K.

As follows from the theory developed in [5], the electric current passing through the system and the accumulated charge are related by the expression

$$Q = C_c U - \sqrt{\frac{C_c L^3 I}{\mu_{eff} d_c}}. \quad (1)$$

It can be seen from expression (1) that, at each instant of time, the accumulated charge is proportional to the square root of the current flowing through the system. Relationship (1) holds for the voltages and times corresponding to the descending branches of the relaxation characteristics. It is worth noting that the lower the voltage and the shorter the time, the better the fit of relationship (1) to the experimental data.

Figure 3 shows the experimental dependences of the charge accumulated in the Pb_3O_4 within different time intervals on the electric current measured at the same times. These dependences were obtained at the applied electric field $E = 8.0 \cdot 10^5$ (curve *1*) and $9.0 \cdot 10^5$ V/m (curve *2*) in the range of the descending branches of the relaxation characteristics. In this case, the currents *I* are the experimental values after subtracting the background current I_b, which corresponds to the leakage current. The background current I_b was chosen equal to the minimum value in the descending branch of the relaxation characteristic at a specified voltage and a longer time. The dependences constructed in the $\sqrt{I}$ –Q coordinates exhibit linear behavior in accordance with the relationship (1).

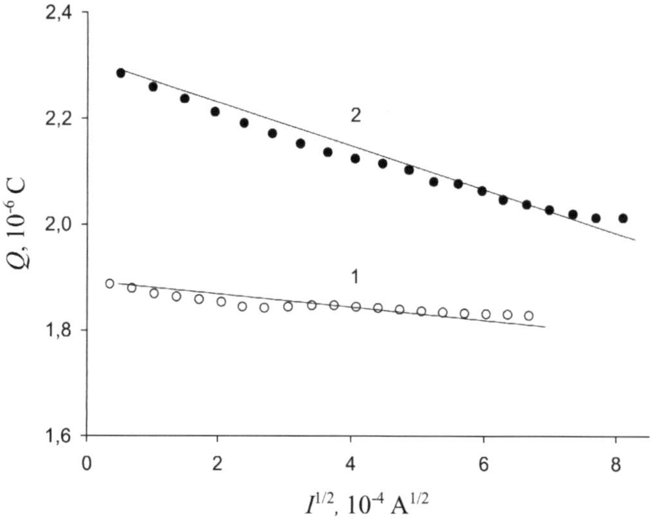

Figure 3. Charge accumulated in the Pb_3O_4 layers for fixed time intervals of the effect of voltage as a function of the current corresponding to the ends of these intervals for the electric fields E: 1 – $8.0 \cdot 10^5$, 2 – $9.0 \cdot 10^5$ V/m. $T = 300$ K.

The curves of current decay for Pb_3O_4 layers at different temperatures are shown in figure 4. It was noted that at high temperatures, the power dependence $I(t)$ goes into exponential. The main contribution to conductivity is probably caused by thermal activation of charge carriers from deep capture levels. In accordance with the relay mechanism, their electron drift mobility μ depends exponentially on temperature [4].

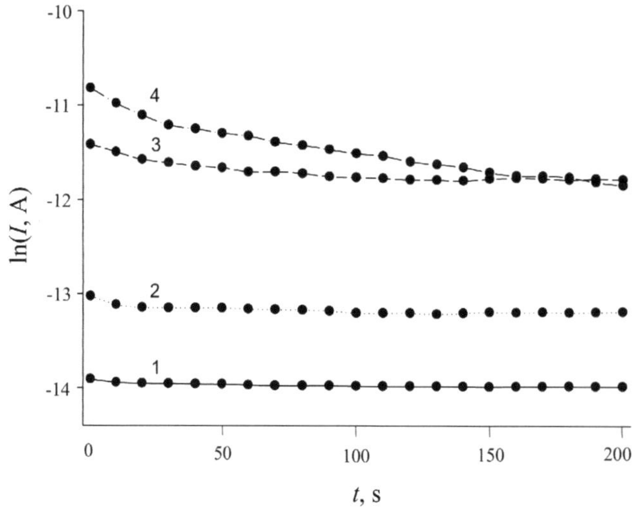

Figure 4. Kinetics of the polarization current measured for Pb_3O_4 layers at temperature T: 1 – 320, 2 – 340, 3 – 360, 4 – 370 K. $E = 5 \cdot 10^5$ V/m.

The exponential decay of current can be determined by the following expression [6]:

$$I(t) = \frac{q\mu\tau n_t SE}{\tau_d}\exp\left(-\frac{t}{\tau_d}\right) \qquad (2)$$

with

$$\tau_d = \frac{1}{N_C\sigma_t}\exp\left(-\frac{E_t}{kT}\right), \qquad (3)$$

where n_t is the initial density of trapped electrons, S – the area that the current is flowing through, E – intensity of applied electric field, σ_t – the conduction of trapped electrons, N_c – the concentration of electrons in the conduction band, E_t – the energy of trapped electrons, k – Boltzmann constant.

According to (2) and (3) the current decays exponentially in time with a time constant equal to τ_d.

Thus, if the measurements are repeated at slightly different temperature, τ_d can be obtained as a function of temperature by plotting ln(I) versus t. Therefore, on the basis of (3), E_t can be deduced from the slope and the intercept, respectively, of the straight line representing $\ln\tau_d$ against $1/kT$.

In figure 5 the plot of $\ln\tau_d$ versus $1/kT$ for Pb_3O_4 is reported. Temperature dependence of $N_c\sigma_t$ has been neglected in comparison to the exponential term of the temperature. The slopes of the dependence $\ln\tau_d\,(10^3/T)$ were determined to be $E_{t1} = 0.13$ eV and $E_{t2} = 0.38$ eV.

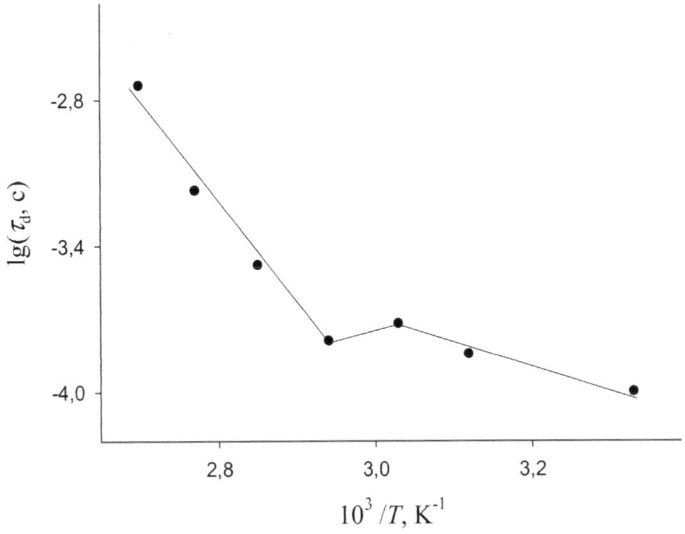

Figure 5. Semi-logarithmic plot of lg τ_d ($1/T$) for Pb_3O_4 layers. $E = 5\cdot10^5$ V/m.

The data on the kinetics of the isothermal relaxation currents were interpreted within a model according to which the transfer of an electric charge injected into a crystal occurs through the relay-race mechanism [2, 3]. The charge transfer from the contact into the sample occurs through the mechanism of hopping conduction over the trap centers, which are located in the band gap and have deep trapping levels of free charge carriers. This process is also accompanied by the formation of an energy barrier at the boundary with the anode, which hinders the transfer of electrons (injected from the cathode) to the anode. The energy barrier at the boundary with the anode arises from the energy difference between the electron affinity for the local center and the work function of the anode metal. Thus, the electric current is limited by both the space charge in the bulk of the semiconductor and the energy barrier at the boundary with the anode, which leads to a redistribution of the voltage across the crystal and decrease of electric current passing through the crystal.

Long-time relaxation processes established for the structures under study on the basis of polycrystalline Pb_3O_4 layers can also be accompanied by the formation of depletion regions at the grain–intergrain layer boundary. A nonuniform potential profile of the boundary regions in the polycrystalline structure studied is the same as that created by a quasi-continuous energy spectrum of the traps.

4. Conclusions

Thus, the behavior of main characteristics of the relaxation process occurring in Pb_3O_4 layers are in agreement with the theory of the relay mechanism of charge transport under the conditions of nonuniform distribution of local centers in metal–(high-resistivity semiconductor)–metal structures. The study of current relaxation related to the accumulation and redistribution of space charge extends our understanding of photoconductivity mechanisms and charge transport in IV–VI compounds with variable ionic–covalent bonding and a tendency towards the formation of naturally disordered structure.

References

[1] Avanesyan V T and Bordovskii V A 2005 *J. Non-Cryst. Solids* **351** 2849
[2] Avanesyan V T, Bordovskii V A and Potachev S A 2002 *J. Non-Cryst. Solids* **305** 136
[3] Mustafayeva S N and Hasanov A I 2004 *Physics of the Solid State* **46** 1937
[4] Timan B L 1973 *Semiconductors* **7** 225
[5] Timan B L 1973 *Semiconductors* **7** 230
[6] Simmons J G and Tam M C 1973 *Phys. Rev.* B **7** 3706

The study of porous silicon powders by capillary condensation

A O Belorus[1], E V Maraeva[1], Yu M Spivak[1] and V A Moshnikov[1,2]

[1] St Petersburg Electrotechnical University "LETI", 5 Prof. Popova str., St. Petersburg 197376, Russia

[2] St. Petersburg Polytechnic University, 29 Polytechnicheskaya str., St.Petersburg 195251, Russia

E-mail: ymkanageeva@yandex.ru

Abstract. Gas adsorption and capillary condensation phenomena were studied in powders of porous silicon with different doping levels of the silicon substrate of *n*-type conductivity in various conditions. Specific surface area and the rated pore size distribution were determined for porous silicon with complex texture by the capillary condensation method. It is shown that the samples obtained in silicon with a higher doping level have a higher value of specific surface area.

1. 1. Introduction

Materials with a high porosity are widely used for sensors, in medicine, as catalysts, radar absorbing, sound and heat insulating coatings for solving problems of component storage, in alternative energy systems of hydrogen storage and so on. Porous silicon (por-Si) and por-Si powders are one of the most promising advanced materials in medicine (targeted drug delivery, implants, photodynamic therapy, biosensors, etc.) [1–4]. The extremely important factors for nanocontainers are geometric characteristics and surface properties of the porous material. The appropriate research methods are required to control the geometric characteristics of the porous silicon and to study the mechanisms of interaction with various substances (conditions of adsorption/desorption, selectivity) with the introduction of one or more drug molecules, fullness, etc.).

In [5] there is an overview of modern methods for the structural study of porous materials. According to [5], the methods for the characterization of the pore structure can be classified according to three basic physical principles they are based on. The first class of methods includes various methods of microscopy providing direct observation of the surface or the cleavage of the studied material. They include scanning electron microscopy, transmission electron microscopy, various techniques of the scanning probe microscopy and others [6]. The second class includes nonpenetrating control methods which are based on the phenomena of scattering of radiation and propagation of waves, positron lifetime spectroscopy [5], and so on. These methods are generally used for the analysis of both open and closed pore systems but require expensive equipment and complex mathematical processing results. Among the third class that includes penetrating analysis methods there are the methods associated with the phenomena of gas adsorption, mercury porosimetry and calorimetric methods.

Adsorption research methods can be considered as the most common way to analyze the material's porous structure. Direct adsorption measurements can be carried out under static conditions, when the transfer of the adsorptive on the solid surface takes place due to diffusion. If the transfer of the adsorptive on surface of the solid is carried out by a flow of the carrier gas, these methods are referred to as flow-through adsorption measurements or dynamic. These methods are widely used in rapid measurements. This investigation method is realized in the devices "Sorbi", in particular, in the device "Sorbi MS" used in this work. The advantage of the method is the possibility to test a sample nondestructively without using toxic substances. The disadvantage of the technique is the possibility to study powders only and using the mathematical models in arriving at parameters of the porous structure.

It should be noted that there are advantages and disadvantages for all of these methods. A full investigation of all sizes of porous is only possible by using several independent methods.

In this work the investigation of porous silicon powders (PSP) obtained by electrochemical etching in n-Si (111) was carried out using gas adsorption and capillary condensation.

2. Experiment

Por-Si powders were obtained in one-chamber electrochemical cell in an aqueous solution of hydrofluoric acid and isopropanol mixture with the addition of hydrogen peroxide [7]. The choice of process conditions (anodization current density of $j = 70$ mA/cm^2, anodization time of $t = 25$ min) was determined by the requirement to obtain mesopores and based on the experience of previous studies [5–7]. As a raw material, the single crystal *n*-Si (111) with a resistivity of 1.0 and 0.3 Ω·cm was used. The por-Si samples thus obtained were then subjected to sonication, and thus yielded a powder of por-Si.

Gas adsorption and capillary condensation measurements were provided by means of SORBI MS ("META", Novosibirsk), software SoftSorbi II, that realizes physical adsorption of noble gas by the sample to be studied. The device SORBI MS allows one to explore the systems with different pore size: to measure the total specific surface according to the method Brauner, Emett, Teller; to measure the external surface area ("external" means without micropores as they are filled with liquid adsorbate) and to evaluate the specific micropore volume; to investigate the processes of capillary condensation in mesopores and to receive information about the distribution of the mesopore sizes.

The features of the method are discussed in [8, 9] concerning the study of the porous sol-gel system "silicon dioxide - tin dioxide". Operating principle of the instrument is based on the comparison of the volume of the gas-adsorbate which is absorbed by the studied sample with the volume of the gas absorbed by the standard sample with a known value of specific surface area. A standard sample was used for calibration of Sorbi. The standard sample has globe-shaped granules (the diameter is 2 mm, the weight is 0.01 g) and its specific surface area is 67 m^2/g. Nitrogen was used as a gas-adsorbate, and helium was used as a carrier gas. For the measurement of absorbed gas volume the thermal desorption method was applied. The specific surface area was calculated according to the model of Brunauer, Emmett, Teller. Before the measurements the samples were heat treated using SorbiPrep station.

3. Results and discussion

In this paper two series of PSP were obtained under the same conditions but with the use of silicon substrates with different resistivity: *n*-Si (111) with 1.0 and 0.3 Ω·cm (series I and II, respectively). Prior to the measurements, a probe of PSP was weighed on an analytical balance (CE124-C, SARTOGOSM). Before gas adsorption measurements, the samples surface was purified from impurities by heating in the sample preparation station SorbiPrep®. Degassing was carried out by heating the sample in a stream of inert gas prior to the determination of its specific surface. Further, the ampoule with the test sample was loaded into the adsorber.

Selection of thermal treatment parameters is important because it affects the accuracy of determining the specific surface area of the sample. If any impurities or the condensed water from the environmental atmosphere remain on the surface, the experimentally determined value of the specific surface area is lower than it should be. On the other hand, the thermal treatment conditions (particularly, temperature) must not alter the structure of the porous silicon. It is known that temperature treatments above 450–500°C in some types of porous silicon could cause the coalescence. Effective degassing conditions will vary depending on the material composition and geometrical properties (pore size, their channel shape, porosity, etc.). The conditions of thermal treatment (thermal treatment time and temperature) were experimentally determined using series I. For series II with higher expected values of the pore diameter the capillary effect is less pronounced, so the adsorbate can be removed easier. Thus, the conditions of thermal treatment chosen for the samples with smaller pores (series I) will be also effective for the samples with larger pores (series II).

To select the suitable conditions for thermal treatment, the specific surface area of the porous silicon powder Type 1 was measured by the gas adsorption method at 4 different combinations of pairs of parameters. The treatment time was 45 and 90 min, and the temperature was 150, 300 and 350°C, and also without any thermal treatment. The results are shown in table 1. According to the measurements of specific surface area for the same sample of series I prepared at different thermal treatment (table 1), the following thermal conditions were selected: t = 90 min, T = 350°C. These conditions were used as well for measurements of samples of series II.

Table 1. The data on the specific surface area for the por-Si powder calculated by the BET method

Series of samples	Specific resistance of silicon wafer, Ω·cm	Mass of sample m, mg	Preparation time t, min	Temperature of the preparation T,°C	Measurement technique; 1 – BET 2 – full adsorption / desorption isotherms	The specific surface area, S, m²/g
I	1.0	10	without treatment	without treatment	1	54.28
		10	45	150	1	49.6
		32	45	300	2	91.63
		32	90	300	1	87.4
		32	90	350	1	104
II	0.3	18	without treatment	without treatment	1	195.2
		18	90	350	1	201

Comparison of the specific surface area S in the samples of different series showed that S in series II (the por-Si powders obtained under the same conditions but in silicon with a lower resistivity) is almost twice higher than S in series I: 104 and 201 m^2/g respectively. Apparently, the dependence of S on the conditions of heat treatment for the samples of series I is stronger due to the presence of smaller pores, the removal of condensate from which requires a higher temperature and a longer exposure time in the thermal treatment station. The samples of series II seem to possess pores of larger diameter, and the condensate layer has less effect on the value of S. A larger specific surface area of the samples of series II results probably from the greater depth of gas penetration (due to larger pore channels), thus involving more volume of porous silicon powder particle into adsorption. In samples with smaller pores (series I) adsorption occurs primarily in the surface layers.

Pore size distribution by capillary condensation was determined for the samples of series I The algorithm for calculating the full adsorption isotherm was carried out using the device SORBI MS.

Figure 1 presents the full adsorption/desorption isotherm with a relative error method, which is 6%. This adsorption/desorption isotherm was analysed with the use of software SoftSorbi2. The results are shown in table 2. As seen from the table, por-Si of the studied type possesses mesopores with the size varying from about 3.5 nm to 56 nm. It should be noticed that there are pores of small diameter (less than 5 nm), the proportion of which was 42.4% of the total pore fraction. Also there are pores with the size of 10–20 nm and 50 nm.

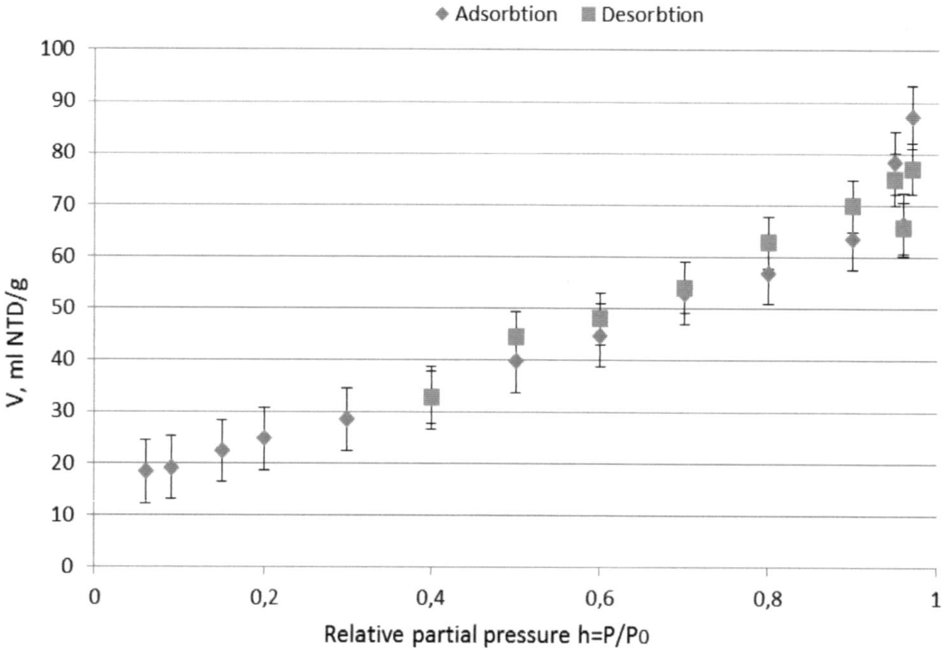

Figure 1. Full adsorption/desorption isotherm for the sample of por-Si formed in *n*-Si (111)

Table 2. Size of pores for the por-Si powders (series I) taken by full adsorption/desorption isotherm

D_i, nm	V_i/V_{total}, %
3.5	34.5
4.4	7.8
5.9	14.2
8.4	19.1
15	5.3
56.1	19.1

4. Conclusions

Thus, the specific surface area of PSP was measured by gas adsorption method. It is shown that under the chosen conditions a decrease in resistance of n-silicon substrate results in a substantial increase in the specific surface area of the porous silicon powder (201 and 104 m^2/g for Si wafer with 0.3 and 1.0 Ω·cm specific resistance respectively). For PSP prepared from silicon (1.0 Ω·cm) the pore size distribution was determined. PSP of series I obtained at the selected conditions of electrochemical anodization possesses mesopores of different sizes: pores with the diameter less than 5 nm (the proportion of which was 42.4% of the total pore fraction) and rather larger mesopores with sizes of 10-20 nm and 50 nm. It was shown that the capillary condensation method enables one to determine the pore sizes in the porous silicon with a complex porous structure (2 or more types of pores, micro- and mesopores).

Acknowledgments

The reported study was supported by the project "Preparation and study of porous systems, functionalized nanomaterials, applications in photonics, sensor technology and medicine" (under the Russian Ministry of Education state task № 16.2112.2014 / K).

References

[1] Belorus A O 2013 *Molodoi Uchenyi, in Russian* **8** 20720297
[2] Lenshin A S, Kashkarov V M, Spivak Yu M, Moshnikov V A 2012 *Materials Chemistry and Physics* **135** 293
[3] Spivak Yu M, Sokolova E N, Petenko O S, Travkin P G 2012 *Molodoi Uchenyi, in Russian* **5** 20720297
[4] Spivak Yu M, Maraeva E V, Belorus A O, Molchanova A V, Nigmadzyanova N R 2014 *Smart Nanocomsite* **4** 115
[5] Maex K, Baklanov M R, Shamiryan D, Iacopi F, Brongersma S H, Yanovitskaya Z S 2003 *J. Appl. Phys* **93** 87938841
[6] Kanageeva Yu M, Savenko V V, Lushinin V V, Moshnikov V A, Zimin S P, Bushin E Yu 2007 *Peterbugskii Jurnal Electroniki, in Russian* **1** 30
[7] Travkin P G, Sokolova E N, Spivak Yu M, Moshnikov V A RF Patent No. 122 385, Bull No. 33 (2012)
[8] Gracheva I E, Moshnikov V A, Karpova S S, Maraeva E V 2011 *Journal of Phisics Conference Sieries* **291** 012017
[9] Gracheva E I, Moshnikov V A, Maraeva E V, Karpova S S, Alexandrova O A, Alekseyev N I, Kuznetsov V V, Olchowik G, Semenov K N, Startseva A V, Sitnikov A V, Olchowik J M 2012 *Journal of Non-Crystalline Solids* **358** 433

AlGaAs/GaAs photodetectors for detection of luminescent light from scintillators

M Bazalevsky[1,2], S Didenko[1], S Yurchuk[1], S Legotin[1], O Rabinovich[1], T Hrustaleva[1], T Suvorova[1] and I Kazakov[2]

[1] National University of Science and Technology "Moscow Institute of Steel and Alloys", P.O. Box 409, Moscow 119313, Russian Federation
[2] P.N. Lebedev Physical Institute of the Russian Academy of Sciences, 53 Leninskiy pr., Moscow 119049, Russian Federation

E-mail: rabinwork@mail.ru

Abstract. The $Al_xGa_{1-x}As$ / GaAs photosensitive heterostructures were grown by molecular beam epitaxy. On their basis the discrete photodiodes (PD) and phototransistors (PT) operating in the visible (0.4–0.7 μm) wavelength range and having photosensitive area diameters of 1.5 mm and 180 μm correspondingly were created. The current-voltage and spectral characteristics of discrete PD and PT were measured and analyzed. Distinctive features of the proposed photodetectors are a high monochromasy with the maximal sensitivity of 0.13 A/W at λ = 530–570 nm and a low dark current of 4.7 nA and 530 pA at 5 V reverse bias for PD and PT correspondingly.

1. Introduction

During the last few years the semiconductor GaAs detectors have attracted a considerable interest as an alternative to the Si and Ge ones for the detection of the luminescent light of scintillators. Photodetectors for the 0.4–0.7 μm spectral range can be used as photosensors in scintillation detectors, which are used, for example, in experiments with cosmic rays. The acquisition of information on the particles falling on the Earth from space is an important research task. However, the fastest accelerators allow getting the particle energies of about 10^{13} eV, while the energy of the particles originating from outer space can reach 10^{17}–10^{19} eV. Besides, due to the sensitivity lack in the infrared range and a high sensitivity in the visible range such photodetectors can be used to control the backlight of liquid crystal displays.

The scintillation detector consists of a scintillator that emits photons when struck by ionizing radiation and a photodetector that converts the light from the scintillator into an electrical signal. In modern detectors optical fibers are used to improve the focusing of the light from the scintillator. Typically, the scintillators used for experiments in high energy physics emit photons at wavelengths from 375 to 430 nm. To detect this light the photomultiplier tubes (PMT) are used because their maximum spectral sensitivity is almost perfectly matches the maximum of the scintillator emission spectrum. However, PMTs have some disadvantages, that is why nowadays researchers and engineers are working on finding an alternative to replace the PMT. The most common alternatives are silicon *p-i-n* diodes [1] or silicon PMTs (SiPM) [2–4].

AlGaAs solid solutions are very promising to create photodetectors for this spectral range. GaAs has the maximum sensitivity at about 800 nm, but a high sensitivity in the visible range (400–500 nm) can be achieved by adding Al that results in an increase of the energy gap. When creating the photodetectors it has been assumed that the scintillation plate of the SC-301 type with the maximum in the emission spectrum at about 420 nm can be used. The wavelength of the light transferred to a photodetector is approximately 476 nm.

This paper presents the results of the investigation carried out with the aim to create high efficiency photodetectors (PDTs) operating in the visible (0.4–0.7 μm) wavelength range. The AlGaAs / GaAs PDTs for detecting the light from the scintillator are presented. Photodiodes and phototransistors based on AlGaAs / GaAs heterostructures (figure 1) were produced using the photolithography techniques with chemically-assisted ion beam etching.

contact layer p^+ GaAs (Be)	$2{\cdot}10^{18}$ cm^{-3}	45 nm
"window" p^+ AlAs (Be)	$2{\cdot}10^{18}$ cm^{-3}	50 nm
p $Al_{0.35}Ga_{0.65}As$ (Be)	$5{\cdot}10^{17}$ cm^{-3}	500 nm
n $Al_{0.35}Ga_{0.65}As$ (Si)	$5{\cdot}10^{17}$ cm^{-3}	500 nm
buffer n^+ GaAs (Si)	$1.5{\cdot}10^{18}$ cm^{-3}	200 nm
substrate n^+ GaAs (Si)	$1.5{\cdot}10^{18}$ cm^{-3}	400 μm

(a)

contact layer n^+ $In_{0.53}Ga_{0.47}As$ (Si)	10^{19} cm^{-3}	40 nm	MBE
contact layer n^+ GaAs (Si)	$4{\cdot}10^{18}$ cm^{-3}	40 nm	MBE
emitter n^+ $Al_{0.25}Ga_{0.75}As$ (Si)	10^{19} cm^{-3}	100 nm	MBE
buffer p GaAs (Be)	$5{\cdot}10^{17}$ cm^{-3}	300 nm	MBE
base p GaAs (C)	$5{\cdot}10^{17}$ cm^{-3}	0.72 μm	MOCVD
i-GaAs	10^{13} cm^{-3}	42 μm	MOCVD
buffer n^+ GaAs (Si)	$7{\cdot}10^{17}$ cm^{-3}	5 μm	MOCVD
substrate n^+ GaAs (Si)	$2{\cdot}10^{18}$ cm^{-3}	500 μm	MOCVD

(b)

Figure 1. Basic structure for photodiodes (a) and phototransistors (b).

Details of the creating and growth for these structures are described in [5–7]. Photodiodes with the circular mesa, the chip size of 4×4 mm^2 and the photosensitive window diameter of 1.5 mm (figure 2 (a)) were mounted by soldering into TO-39 coverage.

Phototransistors were produced in two ways: with and without the base mode. Figure 2 (b) shows a phototransistor chip with size of 600×600 μm^2 which has a round base contact with the bonding pad and the emitter contact; the diameter of the photosensitive window is 180 μm. Phototransistors chips were mounted by soldering into TO-18 coverage. Schematic cross-sections of finally made devices are shown in figure 3.

Journal of Physics: Conference Series **586** (2015) 012018 doi:10.1088/1742-6596/586/1/012018

(a) (b)

Figure 2. Chip view for the photodiode (a) and the phototransistor (b).

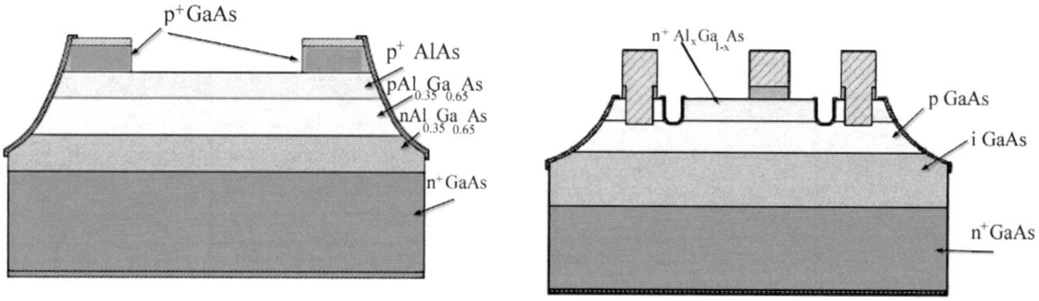

(a) (b)

Figure 3. Cross section for the photodiode (a) and the phototransistor (b).

2. Results and discussion

The dark current and the breakdown voltage were measured by an Agilent B1500A semiconductor device analyzer. The reverse photodiode current-voltage (I–V) characteristic is presented in figure 4. The best samples demonstrated the dark current I_d = 3.38 nA at the reverse bias U_{rev} = 5 V. The I–V characteristic control was carried out at each stage of devices fabrication, from cutting wafers into chips to the packaging process.

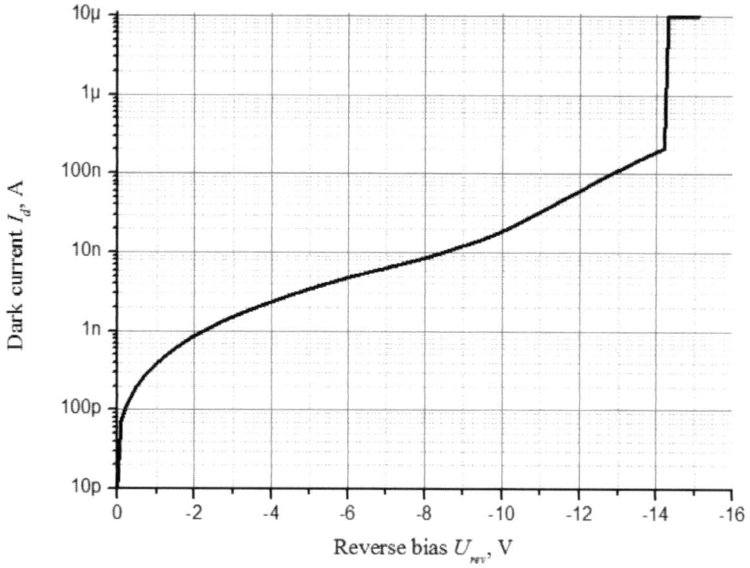

Figure 4. Reverse photodiode I–V characteristic.

The measurements showed that after all production steps the dark current increased and the breakdown voltage decreased insignificantly (figure 5).

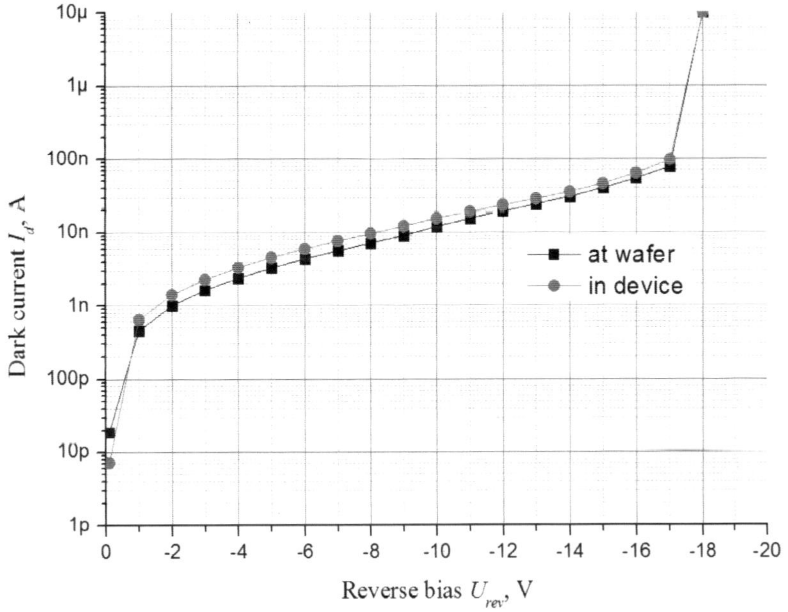

Figure 5. Reverse photodiode I–V characteristics at wafer and after packaging.

The phototransistor I–V characteristics measured at a floating base mode (with two contacts: to the emitter and to the collector) are shown in figure 6. Geometric parameters of samples are as follows: the base thicknesses are 1 µm for 531PT and 0.8 µm for 532PT, the *i*-layer thicknesses are 42 µm for 531PT and 45 µm for 532PT.

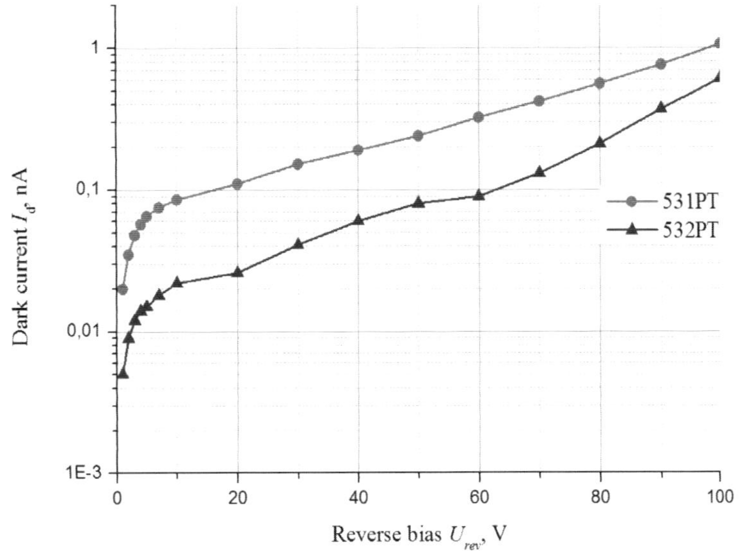

Figure 6. Phototransistors I–V characteristics without base contact for different base thicknesses.

The phototransistor dark currents were 0.1 nA at U_{rev} = 20 V for 531 PT and 0.03 nA for 532 PT; the breakdown voltage was ≈ 600 V. Such small values of the dark current were achieved due to the plasma surface passivation and using a high quality insulating SiO_2. The observed difference between the dark currents can be caused by the fact that the buffer layer in the structure 531 is thicker (300 nm) than in the structure 532 (100 nm), therefore the base layers are of different thicknesses.

Spectral characteristics were measured using a "MDP" type monochromator and a system for measuring the spectral sensitivity of photodetectors TTM 3.435.088. Figure 7 shows the photodiodes spectral characteristics.

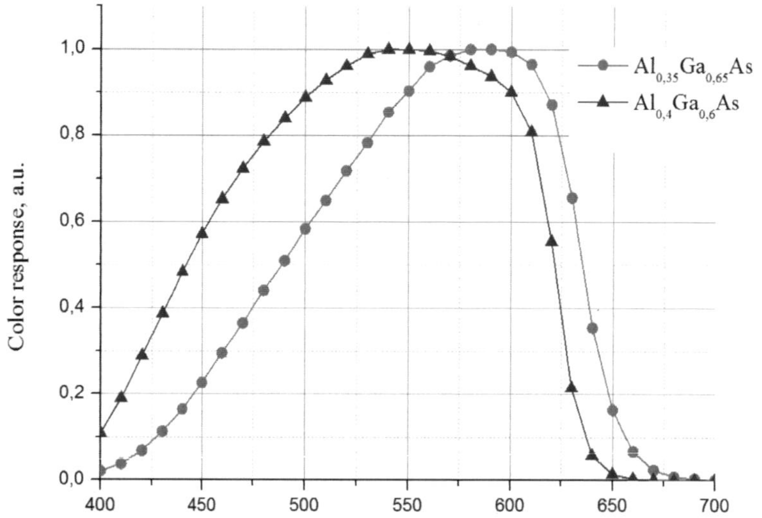

Figure 7. Photodiodes spectral characteristics.

It is seen, that as the Al mole fraction increases by 0.05, the spectral sensitivity maximum shifts by 40 nm to shorter wavelengths. Besides, the sensitivity at a wavelength of 475 nm increases almost twice due to choosing the optimal composition and thickness of the active p-layer. To achieve a greater sensitivity at shorter wavelengths an antireflection coating on the "window" layer can be used. The maximum value of the absolute spectral sensitivity reaches 0.13 A/W.

Unfortunately, we were unable to measure spectral characteristics of phototransistors using the same equipment because the diameter of the incident light beam was 1.5 mm, while the diameter of the transistor photosensitive area was 180 μm. In further experiments such measurements will be made and analyzed.

3. Conclusion

Photodetectors for detecting light from the scintillator based on $Al_xGa_{1-x}As$ / GaAs heterostructures were produced. The performed measurements showed low dark currents at the level of 0.5 nA for phototransistors and 10 nA for photodiodes, as well as high breakdown voltages of about 600 V and 20 V correspondingly. Experimentally determined composition and thickness of layers in the studied heterostructure allowed us to increase the sensitivity at a wavelength of 475 nm almost twice. The absolute spectral sensitivity of photodiodes reaches 0.13 A/W at λ = 570 nm.

Acknowledgments

This work was supported by the Federal Targeted Program "The development of electronic component base and radio electronics" for 2008 – 2015, state contract № 14.430.12.0003 and by the Ministry of Education and Science of the Russian Federation (the base part of the state task "MISiS").

Journal of Physics: Conference Series **586** (2015) 012018 doi:10.1088/1742-6596/586/1/012018

References

[1] Ryzhikov V, Kozin D, Grinyov B, Lysetskaya E, Katrunov K, Kvyatnitskaya V and Chernikov V 2003 *Nuclear Instruments and Methods in Physics Research Section A: Accelerators, Spectrometers, Detectors and Associated Equipment* **505** 58

[2] Herbert D J, D'Ascenzo N, Belcari N, Del Guerra A, Morsani F and Saveliev A 2006 *Nuclear Instruments and Methods in Physics Research Section A: Accelerators, Spectrometers, Detectors and Associated Equipment* **567** 356

[3] Bloser P F *et al* 2014 *Nuclear Instruments and Methods in Physics Research Section A: Accelerators, Spectrometers, Detectors and Associated Equipment* **763** 26

[4] Murashev V N, Legotin S A, Karmanov D E, Baryshnikov F M and Didenko S I 2014 *Journal of Alloys and Compounds* **586** S553–S557

[5] Bazalevsky M A, Koltsov G I, Didenko S I, Yurchuk S Yu, Legotin S A, Rabinovich O I, Murashev V N and Kazakov I P 2014 *Journal of nano- and electronic physics* **6** 03019

[6] Koltsov G I, Didenko S I, Chernykh A V, Chernykh S V, Chubenko A P and Sveshnikov Yu N 2012 *Semiconductors* **46** 1066

[7] Baryshnikov F M, Britvich G I, Chernykh AV, Chernykh S V, Chubenko A P, Didenko S I and Koltsov G I 2012 *AIP Conference Proceedings* **1496** 50

AUTHOR INDEX